A GUIDE TO NEW ZEALAND'S
MARINE RESERVES

JENNY & TONY ENDERBY

First published in 2006 by New Holland Publishers (NZ) Ltd
Auckland • Sydney • London • Cape Town

218 Lake Road, Northcote, Auckland, New Zealand
14 Aquatic Drive, Frenchs Forest, NSW 2086, Australia
86–88 Edgware Road, London W2 2EA, United Kingdom
80 McKenzie Street, Cape Town 8001, South Africa

www.newhollandpublishers.co.nz

Copyright © 2006 in text: Jenny and Tony Enderby
Copyright © 2006 in photography: Jenny and Tony Enderby
Copyright © 2006 New Holland Publishers (NZ) Ltd

Managing editor: Matt Turner
Editor: Jude Fredricsen
Publishing co-ordinator: Dee Murch
Design: Simon Larkin, JAG Graphics
Maps: Chris Edkins (DoC) © Crown Copyright Reserved

ISBN–13: 978 1 86966 114 4
ISBN–10: 1 86966 114 1

National Library of New Zealand Cataloguing-in-Publication Data

Enderby, Tony.
A guide to New Zealand's marine reserves / Tony and Jenny
Enderby.
Includes index.
ISBN 1-86966-114-1
1. Marine parks and reserves—New Zealand—Guidebooks.
2. New Zealand—Guidebooks. I. Enderby, Jenny. II. Title.
333.916410993—dc 22

Colour reproduction by Pica Digital Pte Ltd, Singapore
Printed by Kyodo Printing, Singapore

10 9 8 7 6 5 4 3 2 1

All rights reserved. No part of this publication may be reproduced, stored in a retrieval system, or transmitted in any form or by any means, electronic, mechanical, photocopying, recording or otherwise, without the prior permission of the publishers and copyright holders.

While every care has been taken to ensure the information contained in this book is as accurate as possible, the authors and publishers can accept no responsibility for any loss, injury or inconvenience sustained by any person using the advice contained herein.

CONTENTS

Foreword by Dr Bill Ballantine, MBE	5
Acknowledgements	6
Introduction	7
Map of New Zealand showing marine protected areas	17
Kermadec Islands Marine Reserve	18
Mimiwhangata Marine Park	24
Poor Knights Islands Marine Reserve	30
Cape Rodney–Okakari Point Marine Reserve	38
Tawharanui Marine Park	47
Long Bay–Okura Marine Reserve	54
Motu Manawa (Pollen Island) Marine Reserve	59
Te Matuku Marine Reserve	64
Te Whanganui-a-Hei (Cathedral Cove) Marine Reserve	69
Mayor Island (Tuhua) Marine Reserve	75
Sugar Loaf Islands (Nga Motu) Marine Protected Area	81
Te Tapuwae o Rongokako Marine Reserve	86
Te Angiangi Marine Reserve	91
Kapiti Marine Reserve	96
Westhaven (Te Tai Tapu) Marine Reserve	102
Tonga Island Marine Reserve	107
Horoirangi Marine Reserve	113
Long Island–Kokomohua Marine Reserve	118
Pohatu (Flea Bay) Marine Reserve	123

Fiordland Marine Reserves	128
Piopiotahi Marine Reserve	136
Te Hapua (Sutherland Sound) Marine Reserve	137
Hawea (Clio Rocks) Marine Reserve	138
Kahukura (Gold Arm) Marine Reserve	139
Kutu Parera (Gaer Arm) Marine Reserve	140
Te Awaatu Channel (The Gut) Marine Reserve	141
Taipari Roa (Elizabeth Island) Marine Reserve	143
Moana Uta (Wet Jacket Arm) Marine Reserve	144
Taumoana (Five Fingers Peninsula) Marine Reserve	145
Te Tapuwae o Hua (Long Sound) Marine Reserve	146
Te Wharawhara (Ulva Island) Marine Reserve	147
Auckland Islands (Motu Maha) Marine Reserve	153
Glossary	159
List of species	160
References	167
Further reading	170
Index	172

FOREWORD

This book is a world first. There are many books on marine life – both technical and beautiful. And a few on marine reserves – all very technical. But this is the first book devoted to the range of marine life fully protected in marine reserves.

New Zealand is still a world leader for marine reserves in temperate waters and these reserves cover a wide range of habitats and regions. Jenny and Tony Enderby have explored almost all of them, recording for our pleasure and interest the underwater scenery and the weird and wonderful inhabitants. Many people will be encouraged to visit marine reserves, to snorkel, dive and explore for themselves. Even more, like myself, will enjoy it all through these pages.

Marine reserves are still a new idea, even in New Zealand. Keeping some pieces of the sea free from all human exploitation and disturbance is still considered controversial in some government circles. This is surprising since, as the Enderbys show, once a marine reserve is established, it quickly becomes accepted, popular and successful in many ways. Such reserves are essential for the conservation of the full range of marine life. We have not yet described even half the marine species in New Zealand waters and the only practical way of ensuring their survival is to keep some areas fully protected.

For 25 years I have been saying we should have at least 10 per cent of all habitats in all regions in marine reserves. It is a sad fact that, despite two large reserves around remote islands (the Kermadecs and the Auckland Islands), the 25 marine reserves round the North and South Islands are still just a small scattering.

I believe this will change soon, and I am sure that *A Guide to New Zealand's Marine Reserves* will do a great deal to help it happen. Many politicians are rather confused about marine reserves, most marine managers just stick to their narrow jobs and hope for the best, and most marine scientists only investigate special problems. With the aid of this book, the general public will be able to see that New Zealand's marine life is awesome in its variety and complexity, and that marine reserves are a sensible way to keep it so.

Bill Ballantine, Leigh Marine Laboratory

ACKNOWLEDGEMENTS

Many people from right around New Zealand helped us on our travels and supplied information on New Zealand's marine reserves. Special thanks to Ann McCrone from the Department of Conservation Marine Conservation Unit for answering all our questions and to Chris Edkins for producing the maps. Thanks also to the following people, institutions and operations for their help and tolerance as we rang or emailed: Kevin Bailey of Blenheim Dive Centre, Bill Ballantine, Andrew Baxter, Richard Bray, Barbara Breen, Robin Burleigh, Alan Christie, Russ Cochrane and Wendy Helms from Cathedral Cove Dive, Mark Costello, Debbie Freeman, Roger Grace, Erin Green, Keith Gregor, Jack Hadden, Felicity Heffernan, The Interisland Line, Alan Jones, Vince Kerr, Leigh Marine Laboratory, Margaret Morley, Nelson Dive Club, Fiona Oliphant, Simon Roche, Jason Roxburgh, Tony Roxburgh, Ruth and Lance Shaw of Fiordland Ecology Holidays, Nick Shears, Kala Sivaguru, Megan Stewart, Samara Sutherland, Kathy Walls, Thelma Wilson. Thanks also to Matt Turner of New Holland Publishers and to editor Jude Fredricsen.

INTRODUCTION

Until relatively recently the underwater realm was an alien world, only hinted at by the occasional pod of whales or dolphins that broke the surface or the strange creatures washed up on shores. But over time, humans discovered ways of penetrating the oceans' depths and opening up the hidden realm.

Early research methods were fairly ham-fisted: sea creatures were hauled up by dredges or trawl nets, then handed to scientists for study and identification. Illustrations of dead creatures, no matter how painstakingly drawn by hand, were never going to truly represent the beauty of their living forms. It was not until the 1940s and '50s that the photographs and movies of early scuba-diving pioneers like Jacques Cousteau and Hans Hass began to give the world a true picture of life beneath the ocean waves.

In New Zealand in the late 1940s early divers made their own dive gear and followed in the wake of Cousteau and Hass. Often they were regarded as crazy by their peers, who warned of deadly sharks, moray eels and giant octopus. Most carried spear-guns and hunted the massive schools of kingfish and hapuku that lived in diveable depths around the offshore islands. Returning unscathed from their adventures, they passed on stories of the beauty of the world they were discovering.

New Zealanders began to follow their lead, taking up dive courses run by dive clubs. Advances in scuba technology allowed for more effective and efficient undersea study.

Trouble in paradise

For many years the sea around New Zealand had provided a bountiful harvest of food and beauty to be enjoyed by all. With large pieces of the coastline inaccessible from the land, resident species of fish and shellfish flourished. By the mid 1960s, however, fishers and scientists were already observing drastically reduced numbers in some fish populations. Since then, New Zealand's population has more than doubled. The family clinker dinghy with its 5 hp Seagull outboard motor has been replaced by a seagoing sports craft, equipped with multiple large outboard motors, GPS navigation and electronic fish finders. No reef around New Zealand is exempt from fishing, except within marine reserves or designated protected areas. With the rise in population, the opening up of the coast to the public, pressure from commercial fishing, and the damage caused to coastal and estuarine habitats by silt run-off, trawling, sand mining and other activities, it is no surprise that the country's marine life is under threat.

Attempts have been made to control the impact of these changes, including the implementation of fishing quotas and the establishment of marine reserves.

Fisheries management

Changes to New Zealand's fishing regulations (the Quota Management System) have sought to improve controls over fish taken. Both amateur and commercial catches are regulated by area. Fish stocks are monitored and quotas for targeted species are adjusted accordingly. However, despite these attempts at fisheries management, populations of fish and other targeted marine species continue to fall in unprotected areas. Commercial fisheries admit to harvesting species down to 20% of their unfished numbers.

In addition, the many thousands of non-commercial fish, invertebrates and plants that live in the sea have been poorly served by fisheries regulations. For example, the New Zealand sea horse is protected by law from being traded internationally, but nationally it is up for grabs: anyone can legally collect an unlimited number of sea horses per day.

Marine reserves

Since the 1890s, when New Zealand's first national park was established, almost one-third of the country's land area has been placed under the protection of national park status. It wasn't until the mid 1960s that similar protection was suggested for marine areas: defined zones of coastal sea, protected from fishing and collecting but open to visitors. With bans on fishing and collecting it was hoped that the ecosystems and marine life forms would recover.

A short history
The idea of protecting a piece of New Zealand's coast originated after a marine laboratory was established by Auckland University at Leigh, north of Auckland.

It occurred to Professor Val Chapman that marine studies in the coastal area below the laboratory would be greatly enhanced if pressures on the marine life from collecting and fishing were eliminated. In 1965 he proposed the idea to the Marine Department but was told that there was no legislation to allow total protection for any part of New Zealand's coastal sea, and that they saw no reason to promote protection in any area close to a population centre.

Undiscouraged, Chapman began to champion his idea through public lectures, courses and open days at the Laboratory. He was joined in his crusade by Dr Bill Ballantine, who took up a lecturing position at Auckland University in 1961.

Gradually, support for a marine reserve grew with official backing from the New Zealand Marine Sciences Society and the New Zealand Underwater Association; eventually the Marine Department drafted preliminary legislation and the Marine Reserves Act became law in 1971.

The Act allowed for pieces of the coastal sea to be set aside for scientific research where no fishing, collecting or disturbance of marine life is permitted. It stated that marine reserves should contain 'areas of New Zealand that contain underwater scenery, natural features or marine life, of such distinctive quality, or so typical, or beautiful, or unique, that their continued preservation is in the national interest'. In addition the public would have 'freedom of access and entry to the reserves, so that they may enjoy in full measure the opportunity to study, observe and record marine life in its natural habitat'.

This was a world first: whereas overseas marine protected areas all allowed some fishing, or banned the public from entering the area without a permit, the New Zealand model provided total protection while allowing public access.

Once the Act was passed, the University of Auckland applied for marine reserve status for the coast adjacent to the Leigh laboratory. The Marine Department sought wider consultation on the issue, and received many objections. There was confusion among the general public, many of who assumed they would be excluded from the area. Others feared losing prime fishing spots. Still others pointed out that overseas marine protected areas allowed fishing in some capacity. In 1973 the Marine Department became the Ministry of Agriculture and Fisheries and the decision was held up for two more years. Finally, in November 1975 the application for the creation of New Zealand's first marine reserve, Cape Rodney–Okakari Point Marine Reserve (better known as Goat Island), was approved. In order to fulfil the conditions set out in the Act a management committee was formed in 1976 to erect notices and appoint local rangers, and in May 1977 the reserve was officially opened.

No one knew how quickly the marine life would recover once the area was fully protected. As it turned out, some fish and marine life populations increased relatively quickly, others have taken much longer to recover.

It took many years for the crayfish and snapper in the reserve to grow large enough to eat the larger sea urchins. Over the next 10 years the number of sea urchins decreased, allowing the kelp forest to recover, which in turn encouraged the silver drummer population (a seaweed-eating fish).

Other changes are still happening, with some species returning to the reserve after an absence of over 30 years. Which of these changes are due to the protection afforded by the area's marine reserve status, and which are a natural result of climate and weather patterns is a matter for debate – and further research.

Since the opening of Goat Island, a further 27 marine reserves have been established.

The Marine Reserves Bill is currently before Parliament and when passed will change some parts of the present Act:
- The purpose of creating a marine reserve will change from 'scientific research' to 'conserve indigenous marine biodiversity, in New Zealand's foreshore, internal waters, territorial sea and Exclusive Economic Zone (EEZ) for current and future generations'.
- The fact that marine reserves are no-fishing areas will be made clearer.
- Commercial users will be subject to concessions in the same way that commercial users of national parks are.
- Treaty of Waitangi issues will be included, recognising the Treaty partnership and customary rights of Maori in a proposed marine reserve area and consulting with them at an early stage of proposal.

The challenge for the future

Currently only 5 per cent of New Zealand's coastal sea is fully protected, with the marine reserves around the Kermadec and Auckland Islands accounting for most of that figure.

It is hoped in the future to take a regional approach to creating a network of protected representative marine habitats around New Zealand. The success of the Guardians of Fiordland's Fisheries and Marine Environment (GOFF), a committee formed of representatives of a number of interested parties – everyone from commercial fishers to city-based conservationists – is seen as a model to follow. The comprehensive marine strategy proposed by GOFF saw the creation of eight additional marine reserves in Fiordland, and the protection of 13 per cent of the inner fiords. While marine reserves are not the answer to fisheries management problems, they can offer some valuable insights. By studying species fluctuations along a representative network of marine reserves protected from human influences and comparing them with fished areas, we may gain a better understanding of how best to ensure that future generations will be able to enjoy the produce and the beauty of the sea.

Spin-off benefits from marine reserves

The thousands of visitors to New Zealand's marine reserves are a testimony to the importance of such areas. As are fish like Monkeyface. Monkeyface is a longtime resident of the Cape Rodney–Okakari Point Marine Reserve. Named by visitors for his battle-scarred head, and weighing in at almost 14 kg, Monkeyface has brought enjoyment to thousands of visitors who may never before have seen a live snapper of such a size. Weigh that against the enjoyment of a single smiling angler holding a dead fish on the cover of a fishing magazine.

Conservation

The creation of a marine reserve allows an area of coastal sea to return, as near as possible, to its natural state by protecting all species and habitats. The data shows that marine reserves attract more longer-lived individuals of most locally resident species. Larger animals are generally genetically stronger and less prone to disease. They produce more eggs, and the eggs are more likely to survive and be fertilised. Improved habitats, e.g. healthier kelp forests, mean increased cover for juveniles and more sites for larval settlement. Increased populations lead to some spillover at the edges of reserves, which help to sustain local catch rates. Marine reserves also provide a buffer against fisheries management failures and localised environmental problems.

Research

Marine reserves guarantee scientific studies can be completed without human alteration of the study sites. Protected and unprotected sites can be compared and any differences recorded. It is hoped that further research will reveal whether overfishing, collecting, or silt and nutrient run-off are affecting fish populations or whether such changes are natural. Seasonal variations or migrations can also be identified and monitored. Changes to the biodiversity from an initial baseline study of marine life can be monitored over a long period.

Education

School children, tertiary students and the general public can visit marine reserves to see and study the marine life both in the water and on the rocky shore. Many school groups take the opportunity to compare their local coast with a marine reserve, raising questions about the future of our marine habitats. Marine reserves are the perfect site for underwater naturalist and photography courses.

Recreation and tourism

Marine reserves provide the ultimate destination for international and local recreational scuba divers and snorkellers wanting to see plentiful marine life in a natural environment. Other visitors have the opportunity to look at marine life from the shore, on the rocky reefs or even (as is possible at the Cape Rodney–Okakari Point Marine Reserve) from a glass-bottom boat.

Local businesses, such as recreational and educational centres, food outlets, accommodation and transport providers, benefit financially from the increase in visitor numbers. According to Rodney District Council data, commercial enterprises around the Cape Rodney–Okakari Point Marine Reserve at Leigh, bring more than $20 million to the area each year.

Managing the reserves

Marine reserves are administered by the Department of Conservation (DoC), which was formed in 1987. Prior to this, the duty of care of our marine habitats was shared by the Lands and Survey Department, the Ministry of Agriculture and Fisheries and the Ministry of Transport.

Surveillance and enforcement of marine reserve regulations is carried out by DoC compliance and law enforcement officers and honorary rangers. DoC staff are assisted in this task by other government agencies, including the Ministry of Fisheries, New Zealand Customs and New Zealand Defence Forces.

If you see an offence being committed in a marine reserve you should contact the nearest DoC office or phone the DoC 24-hour hotline, 0800 362 468. For more information on DoC's role, visit their website www.doc.govt.nz.

Marine reserve regulations

A full list of rules can be obtained from DoC. Penalties for offences in marine reserves include confiscation of gear, boats and vehicles, heavy fines and even jail terms. The most important rules are listed below:
- The public can enjoy activities permitted on other public beaches apart from collecting, fishing or damaging the environment. All seashells, washed-up seaweeds or natural materials must be left for future visitors to enjoy.
- No structure is to be erected in or over a marine reserve. This does not prohibit the building of sandcastles or the use of beach umbrellas and shade tents.
- The public must refrain from feeding the fish as it can make them sick, change their natural feeding behaviour and make them aggressive.

Additional guidelines

Boats can move through and anchor in marine reserves but are encouraged to anchor on sandy areas, rather than over reefs. Anchoring on a reef causes major damage to the habitats. Normal maritime safety regulations apply to boats travelling through marine reserves, but extra care must be taken to avoid divers and snorkellers.

It is recommended that boats coming into reserves after fishing outside reserves stow rods and lines to avoid any confusion by other visitors over their intent. Feeding fish with fish scraps or any other food is not permitted. Marine life taken outside marine reserves cannot be released inside marine reserves.

All marine life is protected and nothing should be damaged or broken. While there is no problem observing and photographing marine life such as crayfish, seashells and sea urchins, removing these animals from their habitats, even to look at, can damage and ultimately kill them.

Scuba-divers should be correctly weighted and be aware that loose hanging gauges or poorly placed fins can also damage marine life. Dive knives should never be removed from sheaths apart from in emergency situations.

Other marine protected areas

There are other forms of marine protection as explained below, but none offer the complete protection to the whole ecosystem that marine reserves offer.

Marine parks
A number of marine parks has been established, all of which have different rules administered by the Ministry of Fisheries. Possibly the three best known are the Tawharanui Marine Park in the Hauraki Gulf, Mimiwhangata Marine Park, north of Whangarei, and the Sugar Loaf Islands Marine Protected Area, just off New Plymouth. All three of these parks are featured in this book. While Tawharanui is completely no-take, as are all marine reserves, Mimiwhangata and the Sugar Loaf Islands each have their own limited fishing regulations.

Closed sea mounts
Sea mounts are undersea peaks, some of which rise up to 4 km from the sea floor and reach within 250 m of the ocean surface. Most are volcanic in origin. In the seas surrounding New Zealand over 750 sea mounts have been identified, most falling within New Zealand's EEZ. The largest of these, the Bollons sea mount, near the Chatham Islands, stretches more than 200 km across and 3 km high.

Bottom trawling has reduced fish stocks over many sea mounts and damaged the invertebrate life that lives on them. To protect a representative selection of sea mount habitats, 19 have been closed to trawling.

The marine life around the sea mounts is often unique to a particular sea mount or series of sea mounts. Around 200 species of fish and almost the same number of invertebrates have been identified as living in sea mount habitats; many more species have been observed but have yet to be identified.

Other relevant legislation

The Wildlife Act 1953
This Act was set up for the protection and control of wild animals and birds. The three forms of marine life that come under this Act are: all species of black coral, all species of red coral, and spotted black grouper.

Marine Mammals Protection Act 1978
This Act provides for the protection and management of marine mammals and for the establishment of marine mammal sanctuaries, within which fishing activities can be strictly controlled by the Minister of Conservation.

In 1988 the Banks Peninsula Marine Mammal Sanctuary was created to afford protection to the endangered Hector's dolphin, whose numbers have been decimated by set nets.

The Act was also used to establish a set net ban in 2003 to protect Maui dolphins on the west coast of the North Island, between Maunganui Bluff to Pariokariwa Point, including the entrance to the Manukau Harbour.

In 1993 the Auckland Islands Marine Mammal Sanctuary was set up to protect New Zealand sea lions from being caught in squid trawl nets. (The area is now covered by the Auckland Islands Marine Reserve.)

The Southern Ocean Whale Sanctuary was established by the International Whaling Commission in 1994 and covers all of New Zealand's EEZ south of 40 degrees latitude South.

More details on marine mammal protection can be found on DoC's website, www.doc.govt.nz

Maori protection mechanisms

Mataitai reserves
Under the Fisheries Act 1996, certain traditional fishing grounds, of special significance to local Maori, can be declared mataitai reserves. This means, usually, that no commercial fishing is allowed and recreational fishing is managed by a local Maori committee, who make the bylaws for the area. These must apply equally to all individuals.

According to the Act, a mataitai reserve should not unreasonably affect the ability of local recreational fishers to take fish and should be managed in a way that is consistent with sustainable fishing practices. Laws can be made covering species, quantities, size limits, collecting methods and areas where each species can be taken, all with an eye to maintaining sustainability.

Taiapure

Under the Fisheries Act 1996, local fishing areas of special significance to Maori, either as a traditional food source or for spiritual or cultural reasons, may be declared taiapure. A management committee, which may include both Maori and non-Maori, can recommend the enforcement of fisheries regulations over a taiapure. Anyone may fish in a taiapure. Offence and penalty provisions are similar to amateur and commercial fishing regulations.

Rahui

The Fisheries Act 1996 and the Fisheries Amendment Act 1998 make provision for an area to be declared rahui, which means it is temporarily closed to fishing or that a temporary ban has been placed on certain fishing methods, for the purpose of improving fish stocks and/or because the area is sacred to Maori. A rahui can be applied in a mataitai or taiapure reserve.

About this book

This book covers each of New Zealand's marine reserves plus the three significant marine parks, Mimiwhangata, Tawharanui and Sugar Loaf Islands. The Fiordland reserves are presented as a collective, on account of their distinctive history and ecology. Each reserve shows its location and boundaries on a map and advises you on how to get there. Contact details for the relevant DoC office are given, as they are likely to be your first port of call when it comes to information about the reserve. An abbreviated history for each reserve highlights the main steps along the way to marine reserve status and mentions the groups that lobbied for protection. The marine life sections describe the marine habitats that make up each reserve, and the diversity of marine life that can be seen in each.

The animals and plants living in the sea change rapidly with depth as light and wave action diminish. For each reserve the habitats are divided into zones starting at the high-tide mark and moving to the deeper reefs below the low-tide mark. The high-tide mark is the level the tide reaches only when it is fully in. The low-tide mark is the lowest level the tide reaches when it goes out. The intertidal zone is the area between high and low tides that is exposed when the tide is out. The habitats that are never uncovered by a low tide are known as the subtidal zone.

As experienced scuba-divers the authors offer helpful points for those wanting to snorkel or scuba-dive in the reserves. Where appropriate there is also a brief outline of the various walking tracks that run alongside or around the reserves. Remember to leave time to enjoy the bird and plant life as well as the fish!

Kayaking is another great way of exploring the coast of a marine reserve, and access details are given where suitable.

A word about weather

All visitors to marine reserves need to be aware of the weather and water conditions. If swimming, diving or snorkelling take note of the conditions and if the sea looks too rough, stay out of the water. Dive, swim, snorkel or kayak to your own abilities. If walking close to the edge of the rocks, be aware of wave action. If the rocks are wet it is usually because a larger wave has recently washed over the rocks. These larger waves can occur at any time.

Marine weather forecasts are more comprehensive than normal weather forecasts and give wind direction and changes, sea conditions for the area and tides. Marine weather information can be accessed from VHF Channel 20 and 21, National Radio, daily newspapers, Marine Weather Ph 0900 999 plus area code (i.e. 09 for Auckland area. Note a charge applies), Met Service www.metservice.co.nz/forecasts and Sky Channel 201.

Marine Reserves and other Protected Areas

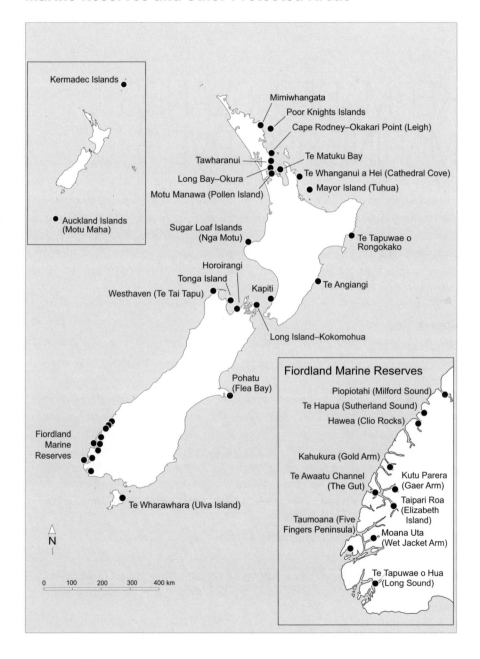

Introduction 17

KERMADEC ISLANDS MARINE RESERVE

These remote islands (annexed to New Zealand in 1887) are the last place on earth where large numbers of spotted black grouper (a heavy-bodied fish of the sea bass family) are found. They form New Zealand's largest marine reserve and are home to tropical, subtropical and temperate marine life, including many species that are found around New Zealand's own coastline. There are many soft and hard corals, but no coral reefs.

Created: 1990

Size: 748,000 ha

Boundaries: Extends 12 nautical miles around each island group (the offshore limit of New Zealand's territorial sea).

Getting there: The Kermadec Islands lie 1000 km north-east of New Zealand, between 29 and 31.5 degrees South and 178 to 179 degrees West. The voyage, usually from Auckland, takes 2–4 days (one way).

Best time to visit: The isolation of the islands and the unpredictable weather make any trip hazardous. Landing and diving can be difficult, particularly during the summer when the islands are visited by tropical cyclones and in winter when violent storms from the south generate huge swells The few commercial trips to the islands usually run in March/April but there is no guarantee of calm weather.

Activities: Snorkelling, scuba-diving, birdwatching.

Facilities: None.

Rules: Marine reserve regulations apply (see page 12) and a Department of Conservation (DoC) permit is required to land. Call the Auckland DoC office on (09) 307 9279 for more information.

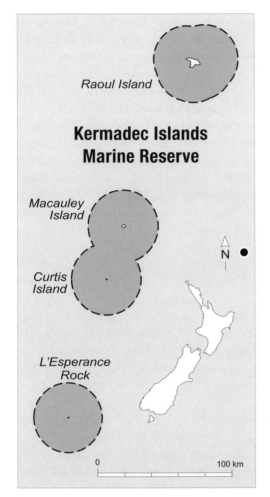

History

Maori legend tells of Rangitahua, an island visited by canoes travelling from Hawaiki to New Zealand. This was probably Raoul Island, the main island of the group known today as the Kermadecs. Carbon dating has put Polynesian settlement of the islands at the 15th century, but they may have been inhabited as early as the 10th century.

Europeans (specifically the captain and crew of the convict fleet escort ship *Lady Penrhyn*) first discovered the islands in 1788 and they were named after French sea captain Huon Kermadec, who visited five years later on the ship *Esperance*.

In the early 19th century, with more than 100 American, British, French and Australian whaling ships working in the area, goats were liberated on Raoul and Macauley islands to provide fresh meat for shipwrecked sailors.

The first European settlers arrived in 1836, but poor soil made raising crops difficult. By the 1860s the sperm whales had been all but exterminated and the whalers moved on. Sometime in the 1870s much of Raoul Island was devastated by volcanic activity and it lay abandoned until 1878 when the Bell family arrived. The Bells lived on the island for 33 years, supplying provisions to the last of the passing whaling ships and exporting timber and fruit to New Zealand.

When the islands were annexed to New Zealand in 1887, Thomas Bell fought the government over ownership of Raoul, and was later granted a parcel of freehold land. A government attempt to colonise the islands failed after several cyclones and an earthquake destroyed buildings and crops, leaving only the ever-resilient Bell family. But even they were finally forced to leave the island in 1911 after a tropical cyclone wrecked their homestead.

In 1934 the islands were designated a nature reserve. An application to use the islands for nuclear testing by the British after World War II was declined by the New Zealand Government in 1947. Interest in protecting the waters around the Kermadecs arose in the 1970s. A change to New Zealand maritime law allowing fishing boats further than 120 km from the mainland meant that the Kermadecs and its fish stocks could be targeted by commercial fishers.

In 1986 the Lands and Survey Department, who were responsible for looking after the islands, made a formal application for marine reserve status, which passed into the hands of DoC when it took over the management of marine reserves in 1987. DoC teams have stayed on Raoul doing weed control and monitoring weather, previously the responsibility of the Ministry of Transport. The islands are being considered for World Heritage Status because of their unique plant and animal life, geological and archaeological values.

Topography

The islands and rocks that make up the Kermadecs are the summits of young, steep-sided volcanoes, caused by the meeting of the Pacific and Australasian plates. Earthquakes are an almost daily occurrence. Most of the islands are pumiceous tuff above andesitic lava. The island chain extends for more than 240 km with the 10,000 m deep Kermadec Trench to the east and the 3000 m deep Havre Trough to the west.

Raoul, the largest island occupies 2938 ha, with its highest point, Mt Moumoukai, reaching 516 m. It is the most active volcanic island in the group, last erupting in 1964, and the only island with a source of fresh water.

To the north-east are the Meyer Islands, Napier Island and the Herald Islets. Macauley Island and Haszard Islet lie 110 km south of Raoul; Curtis and Cheeseman Islands are another 35 km south; and L'Esperance and Havre Rocks, a further 60 km south, are the most southern members of the chain. L'Esperance rises out of 1000 m of water to 69 m above sea level.

Marine life

With water temperatures averaging 18–24°C, the marine life is a mix of temperate and tropical. Fish species number around 150, with the majority tropical or subtropical and only 12 per cent temperate water species. There are soft and hard corals, but no coral reefs. Although a large number of seaweeds are present they do not grow to a large size.

Intertidal zone

The only large area of sand is Denham Bay on the south-west side of Raoul Island, where several species of marine turtle come ashore to lay eggs. There are a couple of small beaches along the northern side, but the large swells which pound the islands almost incessantly make the rocky intertidal zone a difficult habitat with only black nerita snails present in large numbers. Scattered individual barnacles, shore crabs, rock shells, chitons and two small endemic limpets are the only other invertebrates found there.

Three species of Sargassum seaweeds are found on Raoul and Meyer Islands. Large rock pools offer a sheltered habitat to juvenile fishes and tattooed rockskippers (small fish that jump from pool to pool if disturbed).

Shallow zone

At first glance, the shallow zone (low tide to 10 m) appears to consist of bare boulders, devoid of the kelp commonly found in the same zone in New Zealand. However, a closer look at these boulders reveals a covering of small, tough, red seaweeds, providing grazing for endemic Kermadec limpets. These giant limpets, among the world's largest, grow to over 165 mm across. The largest shells are females, and are often attached to homing scars on the rocks that they return to after feeding. Those smaller than 60 mm are usually males and are nearly always attached to the backs of the larger females. Large limpets without males attached are coated with red seaweeds. Between the boulders are brown sea urchins, black Roger's urchins and, less often, pencil and pink urchins.

Warty sea hares and schools of grey drummer are sometimes found feeding on small seaweeds in the shallows. As the name suggests most drummer are grey but one in 10 are bright yellow. Other common seaweed-eating fish are the endemic caramel drummers, bluefish and notchhead marblefish.

Two-spot demoiselles range from the deep reefs to the surface, feeding in the shallow zone. Blue and grey knifefish feed in the white water near the cliffs, as well as trevally and northern kahawai. Yellowtail kingfish are a major predator of the small school fishes and Pacific Gregory – small damselfish up to 15 cm long – are commonly found feeding on seaweeds over the reefs, along with black angelfish and grey Kermadec angelfish.

RESEARCH

Because of the remoteness of the Kermadec Islands, scientific studies are usually restricted to identifying and cataloguing marine species.

Mid subtidal zone

Hard corals appear at depths of 10 m and below, with soft corals appearing at 15 m or so. Although there are reef-building corals, they grow slowly and don't form fringing reefs. The wrasse is the dominant fish, with orange wrasse the most abundant, and Sandager's, green and elegant wrasses reasonably common. Triplefins are scarce with only one species known, although there are several tropical blennies. Pencil urchins with their thick spines are the most common urchin and occasionally crown-of-thorns sea stars are seen, especially at night, when they come out to feed on corals. The hard corals mix with pink coralline algae and some tufting red algaes. Endemic royal top shells, up to 10 cm in height, are the only common grazing gastropods. Sea stars are uncommon except for brown and blue Rodolph's sea stars at Cheeseman Island.

The ultimate experience for the scuba-diver is a close encounter with the large (up to 2 m) spotted black grouper. The Kermadec Islands are now the only place on earth where large numbers of these fish still exist. Had the islands' waters not been protected, the grouper population would have been fished out. Spotted black grouper often inhabit the same hole or cave for many years and divers visiting for a second or subsequent time will often find the same fish in the same spot. They are curious fish and will approach divers, nudging and mouthing them. They even seem to enjoy having their bellies rubbed. Usually solitary creatures, they are occasionally encountered in small schools.

Common northern New Zealand fish include red pigfish, marblefish, painted moki, sharp-nosed puffer, mado, blue maomao and pink maomao. Tropical fish include Moorish idols, yellow trumpetfish, Lord Howe coralfish, long-snouted butterflyfish, black-spotted goatfish and hawkfish. Venomous lionfish hide under ledges during the day but feed at night over the reefs. Cook's scorpionfish sit camouflaged among the coral boulders. Schools of two-spot and Kermadec demoiselles swim together and at night nestle in crevices. Toadstool grouper, gold-ribbon grouper and yellow-banded perch feed at night on shellfish and small fishes. Yellow-striped boarfish form small schools around Raoul. Galapagos sharks, mostly juveniles up to 1 m long, feed over the reefs after dusk.

Deeper reefs and underhangs

At depths of 20 m and below soft corals increase and large, hard, plate corals appear. The soft corals form aggregations over a metre in diameter and completely cover the surfaces of many boulders. Two diadema urchins, a violet-blue species and the spectacular red Palmer's diadema, live at this depth.

Yellow and red gorgonian fans extend from the reef walls, with orange, pink and yellow sponges adding to the colourful display. Other invertebrates include anemones, ascidians and cup corals. Featherstars, yellow and red, hang down, providing shelter for endemic spiny oysters up to 13 cm in diameter. Tiny, tropical brachiopods, only a few millimetres across, are abundant in some areas.

Black coral colonies are common in some places below 25 m and have white, brown or green polyps. Plate corals continue down to below 50 m, mixed with several other species of hard coral.

Marine mammals

Sperm whales, once hunted around the Kermadecs are seen occasionally. Migratory humpback, sei and minke whales pass the islands, and orca, false killer whales and pilot whales also visit. The most common marine mammals are bottlenose dolphins.

Coastal flora and fauna

Endemic pohutukawa and nikau trees dominate the coast. Cats, Norway rats and Polynesian rats took their toll on bird numbers, although all three predators have since been eradicated and the islands now boast some of the densest seabird breeding colonies in New Zealand. The islands are home to a mix of tropical and temperate birds. Masked boobies, red-tailed tropic birds, noddies and ternlets breed along with shearwaters and petrels, some endemic to the Kermadecs. Grey ternlets, which remain at the islands all year round, are the most common birds. Fairy terns nest high in the forks of pohutukawa trees and white-capped noddies nest in small trees on the Herald Islets. Macauley Island has huge populations of seabirds and is the only known breeding locality of black-capped petrels.

Activities

Snorkelling

Charter boats usually have experienced dive guides to identify the best snorkelling sites, depending on sea conditions.

Scuba-diving

Dive guides provided by charter boats will look for the safest dive sites and lead the dives. Large swells and currents can make unguided diving hazardous.

MIMIWHANGATA MARINE PARK

The marine park surrounds a beautiful peninsula that is now a farm park. For most visitors, it's a long drive to get there but worth the effort. The normally clear waters around the peninsula makes marine activities like snorkelling and diving particularly enjoyable.

Created: 1984

Size: 2410 ha

Boundaries: From just west of Paparahi Point, the boundary extends due north out to 1000 m and continues 1000 m offshore, around Rimariki Island and to the south of Te Ruatahi Island.

Getting there: Accessible by boat, car (an hour's drive from Whangarei or Russell) and kayak. From Whangarei take SH 1 north and turn off just before Whakapara. From Helena Bay the road is unsealed and very windy. The nearest boat ramps are at Helena Bay and Teal Bay.

Best time to visit: When there is no easterly swell and winds are light or offshore from the west or south-west. The water is clearer in late summer.

Activities: Snorkelling, scuba-diving, kayaking, swimming, walking, camping.

Facilities: Mimiwhangata Farm Park has toilets, an information board, and camping, lodge and cottage accommodation. For more information, contact Department of Conservation's (DoC) Whangarei office on (09) 430 2133.

Rules: No commercial fishing, no set nets and no long lines. Amateur fishing is permitted using an unweighted, single-hook line or by trolling, spearing or hand-picking. Certain species of fish and shellfish cannot be taken, and normal fishing regulations apply to the allowed species. One pot for rock lobster is permitted per person, party or boat. No dogs are allowed in the farm park.

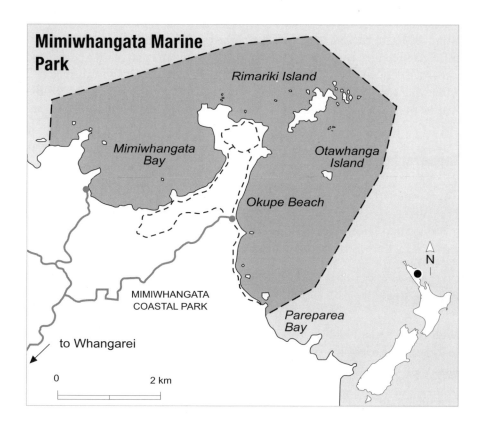

History

The pa sites dotting the high points around the park and the still-visible terraced garden sites are evidence of the long habitation of Maori.

In 1962 New Zealand Breweries (NZB) purchased what was then a working farm with plans to develop the area, but the plans were put on hold when it was realised the area had special ecological value. In 1972 NZB commissioned a report on the peninsula. It recommended that they continue to run the farm, but it encouraged the promotion of natural regeneration and the creation of a marine reserve along the coast adjacent to the farm. In 1975 NZB created a 21-year private trust to run the property as a farm with conservation aims, opening the area to the public as a farm park in 1980. They did not pursue full marine reserve protection as they felt policing 'no fishing' regulations would be difficult and would conflict with their recreational objectives, but they did continue to monitor detrimental environmental and ecological changes. By 1981–82 it was clear that increased visitor numbers were having an adverse effect on some marine life, and in 1984 (around the time the Crown acquired the farm park),

Mimiwhangata Marine Park was established, with a total ban on commercial fishing and limited recreational fishing. In 1987 the farm and marine park came under the control of the Department of Conservation (DoC) and was renamed Mimiwhangata Marine and Coastal Park. In 2002 a proposal to change the marine park to a full marine reserve was raised and public submissions sought.

Topography

The marine park has a mix of sandy beaches, coarse gravel beaches, low greywacke headlands, exposed and sheltered areas. The land adjacent to the marine park is farmed and largely cleared although some wetlands and regenerating forest exist. Numerous small islands and rocks surround the peninsula.

Marine life

With both exposed and sheltered rocky intertidal reefs and sandy beaches, the marine life is extremely varied. Despite the partial protection afforded by marine park status, the marine life differs little from that on the surrounding coast.

Intertidal zone
The rocky reefs each side of the main beach have a series of rock pools containing marine life typical of the exposed coast of Northland. After easterly storms large amounts of seaweeds, seashells and sponges from the deeper reefs wash up along the tideline. Rock oysters have decreased in number. Black nerita snails, cat's eyes, limpets and oyster borers are the most common seashells. Barnacles cover the rocky areas exposed at high tide and provide grazing for tiny blue-and-white periwinkles. Snakeskin chitons sit in the damp clefts in the rocks.

On the sand
Small populations of tuatua and morning star shells live in the shallows. Paddle crabs are common and may have contributed to the reduced numbers of tuatua. Snapper, which would have fed on small paddle crabs in the shallows, are rarely seen thanks to overfishing. Below the low-water mark large dog cockles mix with other bivalve shellfish such as horse mussels and queen scallops.

Gurnard and goatfish dig in the sand and eagle rays and stingrays create much larger holes as they feed on bivalves. Arabic volute shells, up to 15 cm long, with brown lightning patterns, are predators of bivalves. Empty shells are occupied by hermit crabs which peer from under the rim of their temporary homes.

Shallow rocky reefs

Thanks to overgrazing by sea urchins, there is relatively little common kelp growing on the shallow reefs. The loss of predators, namely snapper and red crayfish, has allowed sea urchins to dominate the rocky areas down to depths of around 6 m. Reef fish, such as spotties, banded wrasse, red moki and leatherjackets, are still found in good numbers around the patches of kelp.

Shallow reefs accessible by boat, such as Parrotfish Bay on the east of Rimariki Island, also boast Sandager's, scarlet and green wrasses. Patches of flapjack seaweed cling to the rocks and crested weedfish live almost undetected among its fronds. Curious kelpfish will emerge from the weed and follow divers around. The similar but larger marblefish are also seen occasionally.

Although the reefs are pocked with deep crevices, crayfish are almost non-existent. Most reef walls, especially those on the shaded sides of large boulders, wear a profusion of common and jewel anemones, colonial sponges, hydroids and ascidians. Tiger and trumpet shells are quite common in the cracks and crevices, and large Cook's turban and pink opal top shells graze on the seaweeds. Yellow moray eels are not uncommon with the largest over a metre in length. Black angelfish graze on patches of green sea lettuce. In summer, kingfish feed on schools of piper, blue maomao, sweep, mackerel and koheru. Snapper are rare in the shallows, but more common around the offshore deeper reefs.

Deep reefs

Below 20 m the kelp becomes even more sparse and colonial invertebrate species dominate. Sponges – golfball, yellow, orange and grey encrusting, tall finger – sit among hydroids and ascidians.

At depths of around 30 m the kelp forests vanish and gorgonians (colonial

RESEARCH

Since 1972 Mimiwhangata has been under survey by marine biologists, providing a wealth of information on the effects of limited protection. (Surveys were conducted from 1972 to 1986 and again from 2001 to 2004.) While more than 70 fish species have been recorded – including subtropical species – it would appear that, despite the ban on commercial fishing and the curtailment of recreational fishing, the marine life in the area is not noticeably richer than that of the surrounding coast. In fact, since the first surveys in the 1970s, both fish and shellfish numbers have decreased significantly. Surveys in 1976 showed good kelp forests but by the 1980s these had decreased. Recent surveys commissioned by DoC established that the packhorse crayfish is rarely seen, and red crayfish numbers have stagnated, with few large animals observed during surveys.

corals) begin to appear among the other encrusting life. Red pigfish, gurnard, goatfish and less often giant boarfish feed around the reefs and on the sand. Pink maomao and butterfly perch school over these deep areas feeding on small, drifting plankton. Trevally and kahawai schools move in around late summer although their numbers are not as impressive as they once were.

Spotted black grouper have been seen during several surveys. These members of the sea bass family often remain in the same hole for several seasons. They are similar in size to hapuku, which once lived in the area. Packhorse crayfish were once present in large numbers but are now rarely seen.

Surveys of the deep reefs show an amazing diversity of life, including bamboo corals, black corals and unusual sponges. Large blue moki, foxfish and schools of tarakihi have been filmed during underwater video surveys. It is hoped that this fascinating deep region will return to near its natural state given the full protection of a marine reserve.

Marine mammals

Bottlenose dolphins are regular visitors to the coast where they feed on schooling fish, especially kahawai. Smaller common dolphins pass through the outer area but don't often approach the beaches. Orca (killer whales) pass irregularly and feed on stingrays and eagle rays, often chasing them out of the water onto the beaches. Bryde's whales – coastal baleen whales that grow to 14 m – are also occasionally seen.

Coastal flora and fauna

The park has areas of native bush and a variety of native bush birds including wax eyes, longtail and shining cuckoo. Small numbers of reef herons live on the rocky headlands. New Zealand dotterels and variable oystercatchers nest on the sandy areas, while red-billed and black-backed gulls make their homes around the rocky coast. White-fronted and Caspian terns feed around the coast and are often seen on the beaches. Pied and black shags nest in the pohutukawa trees; pied stilts feed over the tidal flats; and kingfishers sit on fences and dive-bomb invertebrates, both marine and terrestrial.

Little blue penguins come ashore at dusk to nest in the flax and bushes, and Australasian gannets feed in the marine park (although their nearest colony is at the Poor Knights Islands).

Activities

Snorkelling
Usually one side of the peninsula is more protected from wind and swells. Enter from Okupe Beach, Mimiwhangata Beach or Waikahoa Bay. A boat or kayak gives you access to the best snorkelling around the offshore islands and rocks and the northern end of the peninsula.

Scuba-diving
Scuba-diving is easiest from a boat or kayak. There is usually some shelter from wind and swells on the lee side of the islands or peninsula. Beach entry diving is limited to Okupe Beach but swells can make it difficult.

Kayaking
If you've brought a kayak, you can launch it at Helena Bay or Okupe Beach. The 12 km of coastline dotted with rocks and islands, and with many places to land, makes for good kayaking. The camping ground at Waikahoa Bay is just behind the sandy beach allowing for easy kayak access. Contact DoC's Whangarei office (see above) to book a campsite.

Walking
There are well-formed, marked tracks through the farm park. Use tracks and stiles where provided.

Puriri Lookout Track (90 mins return)
The track entrance is marked from Mimiwhangata Road. It's an easy walk through regenerating forest which includes large puriri trees. From the lookout there are views over the forest and coast and out to the Poor Knights Islands.

Tohumoana Track (90 mins return)
The track entrance is marked from Mimiwhangata Road near the parking area. It's an easy walk through farmland and regenerating forest to a lookout with a view over the north-west side of the peninsula and Mimiwhangata Bay.

Peninsula Loop Track (60 mins return)
From the parking area walk to the northern end of Okupe Beach. The track begins just past the lodge and takes you through farmland and past wetland areas, around the end of the peninsula. There is some access to the secluded beaches on the northern end of the peninsula.

POOR KNIGHTS ISLANDS MARINE RESERVE

The clear water and thousands of fish found around these islands give us a picture of what all New Zealand's offshore islands must have been like once. Caves, archways and steep cliffs create unique habitats for colourful invertebrate life. The late Jacques Cousteau rated the Poor Knights as one of the world's top 10 dive sites.

Created: 1981

Size: 2400 ha

Boundaries: Extend 800 m from the main islands, Tawhiti Rahi and Aorangi, and from the Pinnacles (High Peak Rocks), Sugarloaf Rock and all other rocky outcrops and islets above the low-water mark.

Getting there: Accessible by boat only. Tutukaka (30 km north-east of Whangarei) is the setting-off point for most charter boats to the Poor Knights Islands, which lie 24 km to the east. Regular charter boats leave from the marina, where there is a good boat ramp and plenty of parking.

Best time to visit: The water is clearer and warmer in late summer. Although the trip out can be uncomfortable in rough weather, there is almost always a calm anchorage somewhere around the islands.

Activities: Snorkelling, scuba-diving, kayaking, sightseeing, birdwatching, whale and dolphin watching, swimming with the dolphins.

Facilities: None at the Poor Knights Islands, but Tutukaka has toilets, showers, a dive shop, restaurants, motel and bed-and-breakfast accommodation facilities, and campgrounds.

Rules: Marine reserve regulations apply (see page 12) and a Department of Conservation (DoC) permit is required to land. Contact DoC's Whangarei office on (09) 430 2133 for more information. Permits are only issued for scientific study on the islands. Unauthorised landing is not permitted; boats cannot tie up to any point of land.

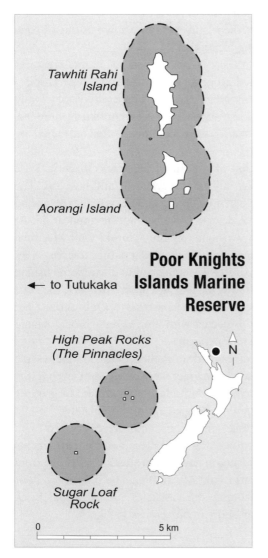

History

Ngatiwai and Ngatitoki hapu (subtribes) lived mainly on Aorangi Island, which had the best canoe landings and gardening sites. Maori oral history tells of an invasion and heavy slaughter of the islands' people, by the Hikutu tribe from the Hokianga around 1820. The islands were never recolonised by Maori and were declared tapu (sacred). In 1882 the Government bought the islands to use as a lighthouse reserve (though the plan to site a lighthouse there never came to fruition).

In 1769 explorer James Cook sailed along the Northland coast and named the islands for their resemblance in his estimation to a breakfast dish known as 'poor knights pudding'.

American author and adventurer Zane Grey was one of the first to publicise the area as a big game fishing destination after making several record catches off the coast in 1925–26. Grey wrote at the time that the Poor Knights Islands would become the most famous of all fishing waters, luring anglers from all over the world, a prediction that would come to pass in a matter of decades.

Pigs, which were introduced by Maori, caused extensive damage to the bush, as well as bird, insect and snail populations, before being eradicated in 1936. All species affected have recovered. Regenerated forest covers the remains of Maori settlement.

In the early 1950s diving pioneer and spear-fisher Leo Ducker and fellow members of the Whangarei Underwater Club discovered the presence of huge fish schools, and from the mid 1960s the islands became known as one of New Zealand's prime dive and spear-fishing locations.

During the late 1960s and early 1970s, large numbers of marine creatures new to New Zealand were discovered around the islands. Some were tropical invaders while others were found to be endemic species. Huge schools of trevally fed on the surface and divers saw kingfish, snapper and hapuku in numbers unheard of today. Bronze whaler sharks were common, often attracted by speared fish.

The first application for the creation of a marine reserve was made in 1972 by the Environmental Defence Society (a group of scientists and lawyers). The NZ Underwater Association already had a voluntary code of conduct regarding spear-fishing in the area and were the driving force behind the application. At the same time, a second application was made by the Hauraki Gulf Maritime Park Board. In 1975 the island grouping was designated a nature reserve.

In 1978, when the Marine Reserves Act was amended to allow some fishing in marine reserves, a third application was lodged, and finally, in 1981, the Poor Knights Islands became New Zealand's second marine reserve. Only about 5 per cent of the marine reserve was fully protected with limited recreational fishing and spear-fishing allowed in the other 95 per cent.

However, it soon became clear to divers and snorkellers that a marine reserve that allowed limited fishing did not work. Snapper were rarely seen, even in the totally protected areas of Maroro Bay and South Harbour. Hapuku or grouper were seen less and less often until sightings stopped almost completely.

In 1994 DoC reviewed the fishing regulations, and were deluged with submissions in favour of total protection. After several years of legal arguments and a second review by DoC, a final decision gave full protection, effective from November 1997, around the southern island of Aorangi. The rest of the Poor Knights Islands Marine Reserve, including the Pinnacles and Sugarloaf Rock became fully protected to 800 m from shore in November 1998.

Topography

There are two main islands, Tawhiti Rahi and Aorangi, and many smaller rocks and islets. The Pinnacles and Sugarloaf Rock are several kilometres to the south.

The islands are the eroded rhyolitic cores of an ancient line of volcanoes which erupted around 11 million years ago. Although connected to the mainland in the past, they have been cut off by the sea for at least 120 000 years. Sea levels during that period have been up to 30 m higher and 100 m lower than today.

Above: *Porae feed during the day on small invertebrates.*

Below: *Palmer's diadema urchins live below 25 m around islands affected by the East Auckland Current.*

Left: *Two-spot demoiselles lay their eggs on rocks and actively defend them from predators.*

Below left: *Red pigfish begin life as females (pictured) but can change into males when they reach around 30 cm in length.*

Bottom left: *Large schools of blue maomao congregate in caves and archways at the Poor Knights Islands.*

Right: *Rikoriko Cave, at the Poor Knights Islands, is the largest sea cave in the Southern Hemisphere.*

Above: *Goat Island shelters the central section of Cape Rodney–Okakari Point Marine Reserve from wind and wave action.*

Left: *Sea horses live right around New Zealand, but their only protection is in marine reserves.*

Above right: *The shore boundaries of marine reserves are indicated by triangle markers, like this one at Cape Rodney.*

Below right: *Groups of goatfish like to rest on sponge-covered rocks below the kelp forest.*

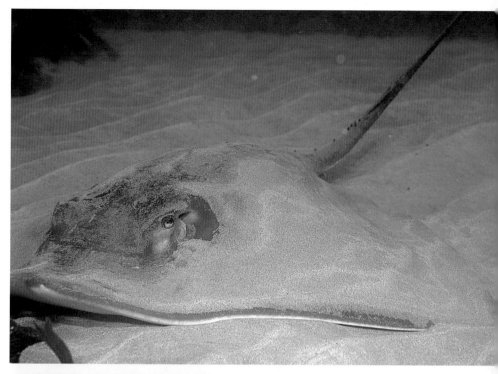

Far left: *A highlight of snorkelling near the beach at Cape Rodney–Okakari Point is being surrounded by a school of blue maomao.*

Above: *Long-tailed stingrays often cover themselves with sand, possibly to hide from predators such as orca.*

Left: *Marine scientists monitor natural changes that occur in marine reserves. For more than 30 years Dr Bill Ballantine (foreground) has studied the intertidal marine life on the reef platform in Cape Rodney–Okakari Point Marine Reserve.*

Snapper numbers in marine reserves like Cape Rodney–Okakari Point are far greater than on surrounding coasts.

On the seaward side of Tawhiti Rahi, the cliffs tower more than 200 m above the water, and continue down for 100 m under the surface. Vertical cliffs are the most common topographical feature, although there are shallow bays at South Harbour and Maroro Bay. Rikoriko Cave, one of the largest sea caves in the world, is big enough for several charter boats to enter. Taravana Cave, with its entrance 20 m below the surface, extends more than 50 m into the island of Tawhiti Rahi.

Marine life

The Poor Knights Islands are in the path of the East Auckland Current that flows from the north along Northland's continental shelf. The current carries tropical and subtropical organisms, usually in larval form, influencing the islands. Most survive only until winter, while others have colonised and are breeding at the islands. The strength and path of the current vary and are usually more influential during La Niña years, when they bring warmer, clearer water. Giant salps (free-swimming marine invertebrates related to the sea squirt), some a metre wide and 10 m long, occasionally appear, along with myriad other marine lifeforms, and more than 120 species of fish have been recorded.

Shallow bays

South Harbour and Maroro Bay are typical sheltered habitats with reefs, boulders and sand. Red Waratah anemones form dense colonies on the shaded reef walls below the low-water mark. Sea lettuce covers the shallow rocks and rimu weed forms thickets down to the sand's edge. Black angelfish are a common sight, feeding on sea lettuce and nesting on rocks. Their blue-and-yellow juveniles, 5–10 cm long, share holes with sea urchins.

Koheru, mackerel and blue maomao feed on plankton above schooling

> **RESEARCH**
>
> After full reserve status was granted in 1998 surveys began comparing fish numbers at the Poor Knights Islands with two reference sites, Cape Brett and the Mokohinau Islands. Surveys were done twice a year for more than six years. Two methods were used: the baited underwater video and the underwater visual census. Results showed snapper had increased dramatically in number and size inside the reserve compared to the two reference sites. Blue maomao, pink maomao, orange wrasse, porae, northern scorpionfish and tarakihi had also increased compared to the reference site.

two-spot demoiselles. Kingfish move through and feed on baitfish.

Snapper have become common but the wrasse is the dominant fish, with Sandager's wrasse the friendliest. The male Sandager's wrasse, resplendent in grey, yellow and purple, is particularly inquisitive and will approach divers and pull at loose-hanging dive gear. Juveniles have a golden lateral line and dark spot at the base of the tail. They act as cleaner fish, removing parasites from black angelfish, demoiselles, porae and goatfish. Bright crimson cleaner fish perform the same service, as do two rarer species, the gaudily patterned rainbow wrasse and the white, black and yellow combfish.

Even on calm days, golden-brown kelp fronds wave gently back and forth. Anemones, small seaweeds, ascidians, sponges and hydroids cover the rocks around the kelp holdfasts.

Yellow-and-blue Verco's nudibranchs (sea slugs) feed on blue-green bryozoans and lay their yellow coils of eggs on them. Bryozoans are colonial animals, but many look like plants.

Long and short-tailed stingrays, some more than 2 m wide, lie under a covering of sand with only their eyes and tails protruding. Two species of lizardfish, red and lilac, both around 30 cm long, wait unseen on the sand for any one of the dozen different species of small fishes known as triplefins to stray too close.

Royal helmet shells move into the shallows during winter, probably to breed. Open-sea visitors that follow the current down from the tropics include the ocean sunfish and several species of turtle.

Vertical cliff walls

Strap kelp grows above common kelp; both sway gently in time with the swells. Triplefins sit on, or dart across, the surface of the rocks and topknot blennies peer out from holes in yellow encrusting sponges. Colourful firebrick sea stars, looking like ceramic creations in orange, yellow and purple, cling to rock faces.

Small red seaweeds live with encrusting invertebrate life around the kelp holdfasts. The spines of large dark Roger's sea urchins point outwards from the rocks and cracks where they are lodged. Scorpionfish rest on rocks, perfectly camouflaged, and red moki, butterfish and leatherjackets move silently around the kelp forests.

Schools of blue maomao give way to pink maomao at 15 m, and below them are splendid perch in pink, yellow and purple. During some summers the schools of two-spot demoiselles include single-spot and yellow demoiselle visitors from the subtropics.

At 25 m the kelp forests begin to thin and sponges become the dominant organisms. Hydroids extend at right angles from the cliffs, like underwater bonsai trees. In their branches are Jason nudibranchs, arguably New Zealand's most attractive sea slugs.

Pairs of Lord Howe coralfish move up and down the cliff walls – the only representative of the tropical coralfish family to have adapted to New Zealand's temperate conditions.

Moray eels occupy the crevices under boulder bottoms and the cracks in the cliff walls. Large yellow morays are the most common but the biggest are the speckled morays, nearly 2 m long. By comparison, the metre-long grey morays seem almost tiny. Mottled moray eels are seen less often.

There are numerous crustacean species, with red crayfish most common, although larger packhorse or green crayfish are occasionally seen. Migrant slipper lobsters are often found on the ceilings of small caves.

Below 20 m the cliffs are often slightly undercut and forests of branching gorgonians (colonial corals) cling to the walls amid more sponges. At 30 m there are typically sand-filled ledges, largely composed of skeletons of the lace bryozoans that line the rock walls.

Caves and archways

The islands' archways are among the most magnificent dive sites anywhere. The Northern, Middle, Barren (now more appropriately known as Splendid) and Blue Maomao archways each have their own special attraction and habitats.

In the Tie-Dye and Northern Archways more than a hundred short-tailed stingrays have been observed moving up and down the walls in summer. Why they come isn't known.

The rock walls inside the archways are a colourful blaze of anemones, sponges, hydroids, ascidians and bryozoans. Tropical banded coral shrimps, with long feelers, are occasionally seen in the cracks.

Porae hang close to the rock walls and small half-banded perch watch from ledges. Hidden in cracks are yellow-banded perch, toadstool and gold-ribbon groupers. Spotted black grouper have begun to reappear. A juvenile pair has taken up residence in Blue Maomao Archway and it is hoped more will return, along with hapuku.

Rikoriko Cave is more barren with a scattering of large boulders across its floor. Mosaic moray eels sit in the cracks on the walls among soft coral colonies and flask sponges. Further back are solitary cup corals and colonies of yellow, anemone-like zoanthids. A night dive here reveals pink maomao and butterfly perch sleeping in crevices while toadstool grouper hunt across the boulders.

Deep reefs

At 35 m and below, light levels are dim. Finger sponges point upwards and golfball and other sponges glow in yellow, orange, pink and white. Pink gorgonians line the edge of the sand and often host spindle cowrie shells that match the patterns of the gorgonians. Gorgonians found on underhangs at lesser depths are prolific here.

The most spectacular sea urchins, brilliant red Palmer's diademas, look black until lit by torch or strobe light. Their needle-like spines are more than 20 cm long and can easily penetrate dive gloves or wetsuits. Schools of two-spot demoiselles, butterfly perch and pink maomao hover and red pigfish follow divers around. Long-finned boarfish have high dorsal fins giving them a distinctly triangular shape. With almost no influence from wave action, the most fragile of animals are able to cling to the cliffs and reefs. At 45 m black coral colonies live among giant sponges and gardens of smaller varieties.

At the Cream Garden, on the north-east side of Tawhiti Rahi, bronze whaler sharks have reappeared around the deep habitats.

Marine mammals

Young male New Zealand fur seals winter over on the Pinnacles and Aorangi Island. Common and bottlenose dolphins are often seen; although it is the larger bottlenose dolphins that are more likely to make contact with divers or snorkellers. Orca (killer whales) make regular visits and feed on stingrays.

Several other species of whale, including humpback, sei and minke, pass close to the islands as they migrate from the tropics to their feeding grounds in the Antarctic. The whale most often seen is the Brydes (pronounced 'broodahs') whale that often feed with gannets and common dolphins between Tutukaka and the Poor Knights Islands.

Coastal flora and fauna

Pohutukawa are the dominant trees and offer a spectacular display in summer with their brilliant red flowers. Poor Knights lilies, only found here and on Hen Island, flower bright red in October and November.

The most common seabirds breeding on the islands are Buller's shearwaters, which are only known to nest here. Every spring more than two million shearwaters return from the Arctic Circle. They feed on the surface around the islands and into the Hauraki Gulf during the day. Other shearwaters include flesh-footed, sooty and fluttering. Petrels include grey-faced, diving, Pycroft's and white-faced storm species. Other seabirds nesting on the islands include little blue penguins, fairy prions, pied shags, white-fronted terns and red-billed gulls. Australasian gannets breed on Sugarloaf Rock and the Pinnacles.

Activities

Snorkelling
The best snorkelling is in Nursery Cove, Maroro Bay and South Harbour as these are shallower areas, where the underwater life can be seen clearly. Charter boat dive masters usually advise on the best sites on the day.

Scuba-diving
There are dive sites to suit all levels of divers. Charter boats usually provide experienced dive guides. This is one of the top dive spots in the world and it's difficult not to be blown away by the colours, fish schools and clear water.

Kayaking
Many charter boats carry sit-on kayaks for their passengers to use free of charge. There are numerous caves and archways that can be explored easily.

CAPE RODNEY–OKAKARI POINT MARINE RESERVE

Snapper, blue maomao and parore abound in the shallows of New Zealand's first marine reserve, better known as Goat Island. Snorkel or scuba-dive over kelp-covered reefs surrounded by fish, and you'll see the benefits of 30 years of marine reserve protection just a 90-minute drive north of Auckland.

Created: 1975

Size: 518 ha

Boundaries: At Cape Rodney two large yellow triangular markers, when lined up, mark the eastern boundary. Two white triangles at Okakari Point mark the western boundary. The offshore boundary is 800 m from the shoreline between Cape Rodney and Okakari Point extending out around Goat Island.

Getting there: The reserve can be accessed by boat, kayak, car or coastal walk. If driving, take SH 1 north, turn off at Warkworth, then follow the Goat Island Marine Reserve signs through Matakana and Leigh. The nearest boat ramp is at Leigh with other ramps at Omaha and Point Wells. Boaties should anchor on sand rather than reef and be particularly careful travelling through the reserve as there are often snorkellers, swimmers and scuba-divers.

Best time to visit: High tide, with less than one metre swell on the east coast and no wind, or an offshore wind from the south-west, are the best diving and snorkelling conditions. Low tide is best for studying the marine life in rock pools and for beach walks. Best underwater visibility is usually in late summer when the water is also at its warmest.

Activities: Snorkelling, scuba-diving, swimming, kayaking, beach and coastal walks.

Facilities: Toilets, changing rooms, cold shower, information kiosk, 5-minute loading zone near beach and parking facilities 100 m up the hill. A glass-bottom boat runs most days from the beach. Dive and snorkel equipment can be hired locally. Shops, restaurants and accommodation around Leigh.

Rules: Marine reserve regulations apply (see page 12). Dogs not permitted on beach. For further information contact Department of Conservation's (DoC) Warkworth office on 09 425 7812.

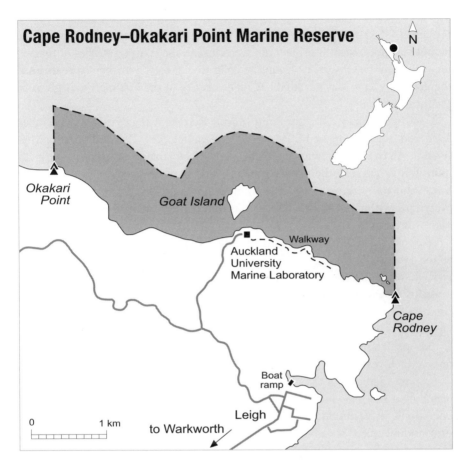

Cape Rodney–Okakari Point Marine Reserve

History

The Cape Rodney–Okakari Point coast has long been inhabited by Maori. A pa was sited on the headland at Okakari Point. Goat Island Bay was known as Whakatuwhenua and the island as Motu Hawere. The coast was a prolific fish, shellfish and crayfish gathering area for local Maori. Ngatiwai are the iwi (tribe) of the area.

Goat Island was named in the sailing ship era, when goats were placed on similar islands for food in case of shipwreck. It is now a scientific reserve.

Stories of large snapper caught from the beach and crayfish collected from rock pools by local farmers date back to the 1940s and earlier. Mountains of kelp were washed onto the beach after storms and these were removed for fertiliser by local farmers using tractors. During the 1950s and '60s the marine life was gradually reduced due to fishing and collecting pressures.

In 1964 Auckland University established the Leigh Marine Laboratory above the cliffs to the east of the beach. In 1965 senior staff at the university thought it would be a good idea to protect the area close to the coast so studies of marine life would not be affected by fishing or collecting. It took a further 10 years until the marine reserve was gazetted. (A fuller history of the first reserve is given in the Introduction.)

The Cape Rodney–Okakari Point Marine Reserve was gazetted in November 1975 and officially opened in May 1977, becoming New Zealand's first marine reserve. At the time it was a typical piece of Northland east coast, with no more fish or crayfish than other coastal area. Some reefs were devoid of kelp forests due to overgrazing by large numbers of sea urchins. From 1975 onwards everything within the marine reserve was totally protected and no fishing of any kind was permitted.

A series of habitat maps covering the whole reserve was drawn up by scientists at the laboratory during 1975–79. The baseline data from these maps have proved invaluable as some changes may otherwise have gone unnoticed.

As there was no reason why people would come to a beach where they couldn't fish or collect seafood, the reserve was not expected to attract visitors. However, as the marine life increased, so did the visitors with numbers reaching 15,000 per annum after 10 years. Today, more than 200,000 people visit the reserve each year.

As part of the flow-on, new businesses were attracted to the Leigh area, including a marine education centre, and glass-bottom boat and scuba-diving operations. Accommodation and food outlets have also increased in the surrounding area.

Topography

The marine reserve is a typical north-east open coast location. Several distinct types of rock are present: ancient greywacke, a very hard rock, best seen on the eastern side of the island; conglomerate material, compressed like natural concrete and containing fossils, to the east of the creek at the beach entrance; and the soft sandstone reefs of Echinoderm Reef and Goat Island Bay. The cliffs behind the bay are layered mudstone, which is eroded by big tides and swells, often discolouring the waters in the bay. A volcanic basalt dyke that runs through the island is visible in the channel opposite the marine laboratory and again on the north-west of the island.

The outer islands of Little and Great Barrier, which lie to the north-east, and Goat Island give the bay protection from wind and wave action.

Marine life

The intertidal reefs of the reserve are typical of the surrounding coast, apart from the increased marine life due to over 30 years of protection. Echinoderm Reef, to the west of the beach entrance, has typical intertidal habitats.

Splash zones and intertidal terraces

The dominant organism on the sandstone terraces is pink calcareous coralline algae. Barnacles cling to rocks in the splash zone, surrounded by periwinkles, limpets and chitons. Small snails such as black neritas and weather-worn oysterborers are scattered or form groups. Dark and white rock shells feed on cat's eyes and top shells. Brown Neptune's necklace seaweed, composed of strands of air-filled sacks, surround the rock pools. If you turn any of the rocks over, the more mobile creatures, like crabs, will race away. Most common are half crabs, with their disproportionately large nippers. Remember to carefully return any rocks back to their original position, or the creatures living underneath will die and be lost to the next group of visitors.

Rock pools

Rock pools often have a dozen different brown, green and red seaweeds. Shrimps dart in every direction and triplefins (small endemic fish) move from rock to rock. Decorator crabs, perfectly disguised with growing seaweeds, wander across the pools. Slow-moving seashells still have the snail that created the shell in residence. The faster-moving snail shells are inhabited by hermit crabs that have taken up residence after the death of the original owner.

Brittle stars, some with thin snake-like arms, others with dark feathery arms, squirm away from under upturned rocks. Related cushion stars vary from orange to dark brown and large predatory spiny sea stars are common. Small sea urchins hide under rocks where predators such as snapper and large sea stars cannot reach them. Colourful colonial animals such as ascidians, sponges and bryozoans colonise the undersides of rocks.

Less often seen are large trumpet shells, often well battered on the outside, each with a brilliant orange inhabitant. In spring several trumpet shell species move onto the reef to lay eggs.

The strangest sea slugs are the 20 cm (or longer) sea hares. On their heads these soft-bodied molluscs have a pair of extended sensory growths known as rhinophores. Sea hares range in colour from brown to red, matching the seaweeds they feed on. Other smaller sea slugs are close relatives. Most often seen are the 3 cm clown nudibranchs with white bodies, orange spots, magenta gills and rhinophores.

A much larger mollusc, the octopus, makes for a fascinating spectacle in a rock pool. Octopus are gentle and inquisitive creatures, and if not threatened, may move out and slide across the pool, changing colour to match the habitat. Red rock and black finger crabs under the larger rocks are the octopus' prey.

Shallow broken reefs

Below the tidal rock platforms, schools of blue maomao mix with snapper and parore. Large snapper, up to a metre long and aged 50 years or more, are the dominant fish over much of the reserve and readily approach divers and snorkellers. Smaller snapper range from 8 cm striped juveniles to schools of 20 cm fish with brilliant blue spots.

Blue cod, clowns of the underwater world, scull in using only pectoral fins. They stop near divers and their eyes swivel independently as they wait and watch.

The brown-and-white striped fish usually seen close to the bottom are red moki that feed on the coralline algae-covered rocks. The thump as the fish tear mouthfuls off with their large rubbery lips can almost be felt. Clouds of sand drift from their gills as they discard non-edible matter. Since the creation of the reserve, red moki numbers and sizes have increased markedly.

Spines of sea urchins protrude from deep cracks between the rocks. Flapjack seaweeds with their tangled fronds attached to a solid stalk are the most common seaweeds. Red crayfish hide under rocks and wait for darkness when they emerge to feed.

A few hundred metres from shore, the number of fish species increases. Jack mackerel, identifiable by their yellow tails, and piper, which look like miniature swordfish, school in the channel in summer. Both species are hunted by kingfish.

Kelp forests

Common kelp with its long stalks and flat fronds, dominates the deeper reefs. Each plant is fixed to the reef with finger-like holdfasts. Only the strongest storms can tear them off, along with the colonies of encrusting animals that live around the holdfasts.

Numerous species of fish have benefited from the increased kelp, using it for food or shelter. Butterfish bite distinctive circular holes in the kelp fronds. Golden juveniles are perfectly camouflaged among the kelp. Male banded wrasse, dark grey with yellow markings on the tops of their bodies, dart in and out of the kelp. Females and juveniles vary in colour from green to brick red. Common spotties school close to the kelp and often hundreds of yellow-finned juveniles swarm close together.

Silver drummer have increased, with large schools throughout the marine reserve. They tear mouthfuls of a variety of seaweeds from the rocks. Their schools often include fish from 20 to 75 cm in length.

Patterned brown kelpfish with domed eyes watch from under the kelp. Much larger, similarly patterned marblefish live in the same area. A dozen species of triplefin range across the rocks and under the kelp.

Metallic grey sweep hover above the weed and sand in mixed schools with juvenile blue maomao, the latter identifiable by their bright blue colour and yellow anal fins.

On the sand

Beneath the sand are myriad worms, crustaceans and molluscs. Goatfish fossick with finger-like barbels under their chins. Often spotties or small snapper wait for the goatfish to uncover a worm or crab then compete for the meal on offer.

Trevally and snapper leave holes in the sand as they feed. Larger holes are created by eagle rays and shining, broken pieces of Cook's turban shells are

RESEARCH

The underwater habitat maps created by Leigh Marine Laboratory staff in the late 1970s showed large areas where sea urchins were abundant and kelp almost non-existent. These areas, known as urchin barrens, were thought to be a normally occurring phenomenon. However, between 1978 and 2000 the barrens have reverted to kelp forests and mixed seaweeds as the crayfish and snapper (major predators of the sea urchins) have recovered.

In 2003 the Leigh Marine Laboratory conducted surveys using baited underwater video within three marine protected areas – Cape Rodney–Okakari Point Marine Reserve, Tawharanui Marine Park and Te Whanganui a Hei Marine Reserve – in order to measure snapper numbers. The three, on average, held 14 times the number of snapper compared to similar unprotected habitats surveyed. In the Cape Rodney–Okakari Point Marine Reserve legal-sized snapper were twice as abundant than in the previous autumn's survey.

Red crayfish have been studied over 25 years and considerable data compiled. The numbers of crayfish within the reserve varies greatly, with the irregular peaks and troughs in their population reflected in commercial crayfish catches on the surrounding coast. In 2002 a study compared catches on the reserve border with those on the Leigh coast and Little Barrier Island. The authors of the study concluded that the number of crayfish leaving the reserve indicated that catch rates on the reserve boundaries were similar to levels on nearby coasts. This in turn meant that local fishers were not necessarily disadvantaged by the establishment of a marine reserve.

evidence of a successful meal. Long and short-tailed stingrays are usually seen further out in the bay. When resting the stingrays use their wings to cover themselves with sand and only the outline of their bodies, eyes and tails give them away.

Silver drummer and parore hover at cleaning stations, their bodies changing to almost pure white inviting tiny trevally, around 5 cm long, to remove parasites.

Giant boarfish were once only seen at depths of 20 m or more. These brown-and-white patterned fishes, half a metre long with extended snouts, now regularly feed not far from the beach.

Deeper reefs and sponge gardens

North Reef, off the island's north-west point, comprises a series of pinnacles, rising from a depth of more than 10 m to almost break the surface. Its base is typical of deeper reefs in the reserve, with rocky walls covered in encrusting marine animals. Common and jewel anemones resemble garden flowers, hydroids look like white bonsai trees, and colonial sponges and ascidians add patches of colour.

Large numbers of red crayfish jostle for space at the rear of cracks and caves, the largest more than 3 kg in weight. Occasionally they share their space with the rarer packhorse crayfish or slipper lobsters.

Yellow moray and conger eels also live here. Further back in the caves are night fishes, slender roughie and big eye, all of which emerge at dusk to feed in the open.

Leatherjackets, New Zealand's only triggerfish, congregate over the kelp and huge schools of silver drummer race in and away. The dull red of the pigfish contrasts with the schools of sweep and blue maomao. Sandager's, green and scarlet wrasses are seen irregularly. Kingfish move through in summer, picking off stragglers from the mackerel and koheru schools.

The kelp forests become scarce at 15 m as the sunlight they need to grow is reduced. Sponges now dominate the sea floor. Grey, yellow and orange finger sponges tower above golf ball, grey pillow and encrusting varieties.

Schools of two-spot demoiselles and butterfly perch hover above the sponges, feeding on drifting plankton.

Marine mammals

Bottlenose dolphins are the most common marine mammals. Pods of 5–20 animals regularly feed close to the beach on fish schools, especially kahawai. Orca (killer whales) visit less often and feed on stingrays and eagle rays. Smaller cream and grey common dolphins feed at the edge of the reserve and in early summer Bryde's whales move into the plankton-rich waters to feed.

Coastal flora and fauna

Pied shags nest in the large pohutukawa trees above the beach and on the island. After feeding they stand on the beach or rocks, with wings extended, to dry. Introduced mallard ducks and less often paradise ducks wander the beach especially near the car park and stream. Along the beach kingfishers, fantails, New Zealand pipits and yellowhammers are a common sight. Flaxes and invasive pine trees, gorse and pampas grass dominate the hillsides.

Little blue penguins feed at sea during the day and return to the beach and island at night to nest among the trees and flax. Reef and white-faced herons, oystercatchers, petrels, white-fronted and Caspian terns, red-billed and black-backed seagulls feed on the tidal reefs and some nest on the island and secluded parts of the coast. Australasian gannets, although not resident, often feed in the reserve.

Activities

Snorkelling
There is good snorkelling close to the beach and rocks. You can snorkel easily to the island or Shag Rock not far from shore. The schools of fish make great company in a safe snorkelling environment.

Scuba-diving
Park in the 5-minute zone to unload, then move your car to the main car park. It's a short walk across the sand and into the water. Don't enter or exit across the rocks as they can be slippery. Listen for moving boats before surfacing.

Kayaking
Kayaks can be unloaded in the 5-minute loading zone and launched and retrieved from the beach. The island is easily circumnavigated and has some large caves, but do not enter if any swell is running. It is also possible to launch your kayak at Leigh Harbour or Matheson Bay and the trip around Cape Rodney is an easy 60–90 minute paddle. Conditions at Cape Rodney can be rough when the wind opposes the tide or swell. Kayaks can also land at Pink and Pakiri Beaches but most of the coast is rocky and unsuitable for landing.

Walking
Beach walk (2–3 hrs return)
At low tide you can walk west along the beach and rocky shore to Pakiri Beach. Allow two hours return to Okakari Point or three hours return to Pakiri Beach.

The sandy beach changes to boulders, then sandstone rock platforms. Near the start of the rock platform are the remains of a concrete and rock structure, known as the 'Niagara Wall'. It points to where the passenger liner *Niagara* sank near the Hen and Chickens Islands after hitting a mine in World War II. Just before Pink Beach, about halfway to Okakari Point, there is a rock pool with large numbers of juvenile parore. The pinky sand of Pink Beach gets its colour from millions of broken wheel shells. From Pink Beach the walk continues under steep cliffs, across rocky intertidal platforms to Okakari Point. Be wary of falling rocks at all times.

Coast walk (2 hrs return)
The track begins behind the Leigh Marine Laboratory and is clearly marked with posts. It crosses farmland then follows the coast. Care is needed as the track passes close to the cliff edge. The track detours inland through native bush and finishes overlooking Tabletop Reef. Take a seat and enjoy the stunning views out to Little Barrier and the Hen and Chickens Islands. The track is very slippery after heavy rain.

Cliff walk (1 hr return)
An unmarked track follows the cliff edge, west from the car park. Take care as it is largely unformed and parts are affected by slips. It offers great views over the reserve and out to the Hen and Chickens Islands.

Note: Park in the public car park for all walks.

TAWHARANUI MARINE PARK

Huge crayfish are a feature of this marine park on the northern side of the Tawharanui Regional Park. No other marine park has shown an equivalent increase in marine life. Thanks to the total 'no-take' regulations the prolific underwater life complements the now predator-free regional park, which is being replanted and thus attracting endangered birds.

Created: 1981

Size: 588 ha

Boundaries: The eastern and western ends of the marine reserve are marked by white and orange triangular markers. When the markers are aligned vertically from sea, they mark the boundary, which extends to half a nautical mile from shore. The eastern boundary extends north from a headland west of Takatu Point. The western boundary extends northeast from east of Pukenihinihi Point.

Getting there: Access is by boat, car, kayak or coastal walk. The Tawharanui Regional Park is located on the Tawharanui Peninsula, 90 km north of Auckland. If driving, take SH 1 north of Auckland to Warkworth. Turn off onto the Leigh Road until you get to the Omaha/Tawharanui intersection. Turn right, and then right again at the roundabout. The park is well signposted from Warkworth. The nearest boat ramps are at Omaha, Point Wells, Leigh and Sandspit.

Best time to visit: Low tide is the best time to explore the rocky reefs. Scuba-diving, surfing and kayaking can be enjoyed year-round, although you will need to check sea conditions before setting out.

Activities: Snorkelling, scuba-diving, swimming, surfing, kayaking, walking.

Facilities: Toilets, changing rooms and information kiosk in the regional park.

Rules: A total 'no-take' ban, i.e. no fishing, damaging or collecting marine life. No dogs allowed in the regional park. Contact the Auckland Regional Council 0800 80 60 40. If you see an offence report it to the Ministry of Fisheries 0800 478 537.

Tawharanui Marine Park

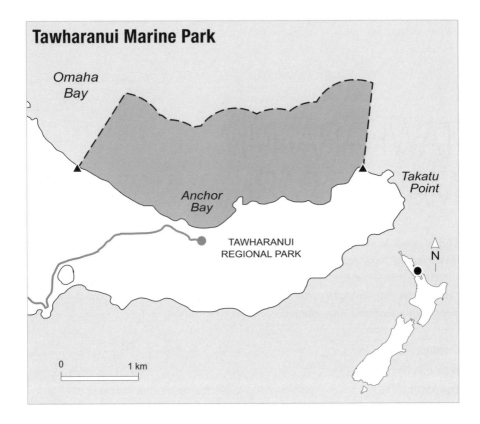

History

Maori occupation of Tawharanui ('the abundant edible bracts of the kiekie vine') can be traced back a thousand years. Te Kawerau and Ngati Raupo were the main tribes.

Phoenix Reef and Comet Rocks were both named after ships wrecked there. Early European settlers milled kauri and puriri, and felled manuka for firewood. In 1853 the Kawerau iwi (tribe) sold the land to the Crown, and in 1973 the Auckland Regional Authority (ARA) purchased it and established Tawharanui Regional Park, providing public road access to the peninsula for the first time. When a survey showed that the marine life along the northern coast was of particular interest, the ARA suggested the creation of a marine protected area next to the park where visitors could enjoy and study the marine life. However, the Marine Reserves Act 1971 did not provide for the kind of recreational marine park the ARA had in mind, and they had to make do with a grant of control and fishery protection under the Harbours Act 1950. This meant control of most activities except fishing. Finally, in 1981 an amendment to Fisheries Regulations provided that 'no person shall take any fish from the area to half a nautical mile from shore' and the marine park was established. While these regulations provide an element of protection, the present fishing rules can be altered or rescinded at any time by ministerial decision. In order to bring the marine park into line with other areas protected under the Marine Reserves Act 1971, in February 2003 the Auckland Regional Council (ARC, which replaced the ARA) proposed changing the marine park to a marine reserve, under which conditions DoC would govern compliance and law enforcement and management.

Topography

The rocks along Ocean Beach are part of the Waitemata group: yellow sandstone, conglomerates, grits and siltstones. The soft rocks at the east end of Anchor Bay contain rich fossil deposits. Slumping of sediment on the sea floor 20 million years ago has produced the folded swirls of Comet Rocks. Over the last 5000 years, sand dunes have built up behind the main beach.

Marine life

Tawharanui's coast is a mix of open sandy beaches and kelp-covered sandstone reefs. Although there is little run-off from the land adjacent to the park, a lack of currents in Omaha Bay leaves fine particles of silt.

Intertidal zones

The marine habitat changes from open sandy beach to low tidal reefs to the deep waters of the bay. From Anchor Bay to the eastern boundary of the park, rocky

RESEARCH

Marine life has been monitored irregularly since 1977 and in recent years surveys have shown large numbers of very big crayfish within the protected area. The increase in crayfish numbers was most dramatic in surveys conducted through 2004–05. In fact, a survey in 2005 showed a crayfish biomass 25 times greater than in similar habitat sites outside the marine park.

Surveys of snapper numbers, conducted by the Leigh Marine Laboratory, also show higher numbers than the surrounding coast, but not as high as the Cape Rodney–Okakari Point Marine Reserve. Some illegal fishing may account for the lesser numbers.

reefs and small pebbly beaches are prevalent. Numerous brown, red and green seaweeds cover the rocks, including bloodweed and grapeweed.

Oysters, rockborers and cat's eyes are the most common shellfish. Under the rocks are huge numbers of half crabs, easily identified by their very large pincers. Close inspection of the rocks reveals small sea urchins and cushion stars. Several species of chiton cling tightly to the undersides of rocks.

Rock pools

As the tide retreats very large rock pools appear around Anchor Bay and the adjacent Comet Rocks and Phoenix Reef. Walk across these rocky areas and you will quickly become aware of the abundant marine life. Large numbers of grazing seashells such as cat's eyes, Cook's turbans and spotted top shells share the pools with sea stars, sea urchins, hermit crabs and red rock crabs.

In the deeper pools near the low tide mark, an eye peering from a hole or an orange-lined tentacle with suckers gives away the octopus, which can be sighted all around New Zealand's coasts. Numerous species of brown, red and green seaweeds provide cover for grazing sea shells, shrimps and small fish. Several varieties of triplefin – small fish common to rock pools – and common shrimps dart around. What look like patches of moving weed are actually decorator crabs, which snip pieces of living seaweed and attach them to their backs. Occasionally kelpfish, trapped by the falling tide, sit perfectly camouflaged amongst the seaweeds, but dart away if disturbed.

Near the low tide mark among the seaweeds are sea hares – distinctive shell-less molluscs – and the related sea slugs or nudibranchs. Lemon nudibranchs, around 4 cm long, are the most common.

Shallow broken reefs

Below the low tide mark, shallow reefs support large populations of adult sea urchins that graze on the seaweeds. The sandy areas at around 6 m depth, off Anchor Bay and to the west, are interspersed with reefs where snapper, spotties, leatherjackets, red moki and banded wrasse are common. Some schools of silver drummer have appeared in the past few years and feed around the edges of the kelp forests.

Large concentrations of crayfish occur right along the rocky reefs. The shallow, subtidal reefs between 5 and 15 m depth, seasonally play host to hundreds of animals. Red crayfish are abundant but there are also small groups of green or packhorse crayfish. While the red crayfish tend to be large individuals, weighing up to 4–5 kg, the packhorse crayfish tend to be juveniles. The reefs change to sand at around 5 m although scattered patches of reefs continue out to more than 20 m.

Kelp forests

Common kelp is continuous along the reefs from Anchor Bay to the eastern boundary down to around 20 m depth. The underhangs and shaded walls beneath the kelp hide a wealth of colourful colonial ascidians, sponges, hydroids and bryozoans, interspersed with patches of common and jewel anemones.

Cook's turban, trumpet, whelk and tiger shells nestle in the rock cracks and crevices and among the kelp holdfasts. Of the nudibranchs, orange-spotted clown nudibranchs are most likely to be seen, while blue-spotted gem and brown-and-white variable nudibranchs, up to 7 cm long, are harder to find despite being relatively common.

Crayfish numbers are high and many larger animals venture out in the open in daylight, away from the cracks in the reefs where they usually congregate. A few large snapper, over 6 kg, stand out among the many smaller individuals. Schools of blue maomao and sweep feed above the kelp forests.

On the sand

Like much of the coast, the sandy areas below 15 m depth have been colonised by huge numbers of parchment worms which may prove to be invasive. The effect of this animal on the resident seabed populations is as yet unknown. The worms' 7–12 cm white, parchment-like tubes wash up after onshore winds.

Eagle rays and long and short-tailed stingrays rest on the sea floor under a thin coating of sand. Broken Cook's turban and scallop shells are evidence of their feeding.

Deeper sandy areas support fish like gurnard and giant boarfish (not often seen by divers) that feed on worms and crustaceans buried in the sand. Scallops appear in small numbers, sharing the sand with large-headed stargazers that sit with only their eyes and mouths exposed and guitar-shaped electric rays.

Deeper reefs

At the eastern end of the marine park, deep canyons drop to more than 20 m and the reef walls are home to a colourful array of invertebrate life. Hydroids, ascidians, anemones and sponges in many varieties cover the rock faces and nudibranchs graze over them. Two endemic nudibranchs are usually found near their preferred food source: Jason nudibranchs on white hydroids and apricot nudibranchs on dead man's finger soft corals.

On these deeper reefs, scarlet wrasse, John dory and dwarf scorpionfish are quite common. Schools of tarakihi, trevally and kahawai appear seasonally. Mado, 15 cm-long fish with horizontal yellow, black and white stripes, appear in small schools and are scarce on the surrounding coast. Sponges like the golf ball and long grey finger sponges grow in sheltered gullies between the reefs.

Marine mammals

Bottlenose dolphins are regular visitors to the reserve and often feed and play for hours. Orca (killer whales) move through infrequently and feed on eagle rays and stingrays. Bryde's whales feed in the Hauraki Gulf almost all year round and often move close to the beach and reefs around October to December. Occasionally, their spouts and backs – with their distinctive sickle-shaped dorsal fins – are seen from the cliffs to the west of the beach.

Coastal flora and fauna

The regional park is being replanted in native trees and shrubs and wetlands will be re-established. The park was cleared of most mammalian predators in 2004 and a predator fence erected. Native species now extinct on the northern New Zealand mainland will be reintroduced to the park, although bellbirds have already established naturally. Small numbers of New Zealand dotterels nest on the dunes and oystercatchers are common. Little blue penguins, grey-faced petrels, white-fronted terns and seagulls all feed in the marine park.

Activities

Snorkelling

All the reefs within walking distance of the car park offer good snorkelling. Best access is from the sand, rather than walking across the slippery rocks when entering the water.

Scuba-diving

Because of the distance from the car park, scuba-diving is easiest off a boat or kayak. There are many reefs dotted along the coast that offer good diving to suit all skill levels.

Surfing

Anchor Bay offers good surfing between the reefs especially when a large swell is running and the wind is offshore.

Kayaking

Kayaks can be launched at Jones Bay at the entrance to the park. It's about a 10 km paddle around the peninsula. Sea conditions can be very rough at Takatu Point. If the swell is small enough you can land or launch on Ocean Beach. Another option is to set off from the southern end of Omaha Bay or Whangateau Harbour and follow the coastline to the park.

Walking

The park has a range of walks to accommodate all levels of fitness.

South Coast Walk (2–3 hrs return)

This coastal walk starts from the small car park near the lagoon and ends at Takatu Point at the eastern end of the peninsula. There are excellent views over the Hauraki Gulf and to Little Barrier Island. Allow 3 hours return. A shorter version starts from Anchor Bay car park along the ridge to Takatu Point and takes around 2 hours return.

Ecology Trail (90 mins return)

The walk begins at Anchor Bay and follows the beach to the east passing several large caves that can be explored at low tide. At the end of the beach it climbs to the ridge with good views over the marine park. Climb over a stile and follow the track through a section of regenerating native bush with identification posts for plants. Alternatively follow the ridge track back to Anchor Bay. Allow half an hour return for the shorter version.

Tawharanui Beach Walk (1 hr return)

From the car park access to Tawharanui Beach walk west along the beach. Best time is at low tide. This easy beach walk covers a distance of about 2 km each way and takes you to where the predator fence joins the beach. Be aware of bird nesting sites if you are walking near sand dunes. At low tide you can walk across the exposed rocky platform near Comet Rocks.

LONG BAY–OKURA MARINE RESERVE

Long Bay's sandy beach and rocky shore are typical of northern Auckland's sheltered coasts. Its diversity is added to by the Okura River with its sand flats and muddy mangroves and which provides a sheltered habitat for a variety of juvenile fish. The marine reserve is adjacent to the Long Bay Regional Park.

Created: 1995

Size: 980 ha

Boundaries: The southern boundary at Toroa Point is marked by two yellow triangular markers. It extends from MHWS (the high water mark on the highest tides) northeast 0.9 of a nautical mile to a red marker buoy. The boundary then travels approximately northwest – in line with a series of buoys – to a red navigational beacon at Karepiro Bay, and west to two yellow triangular markers. The reserve includes the Okura River Estuary.

Getting there: The reserve adjoins the Long Bay Regional Park, just 22 km north of Auckland. Access is possible by car, boat, kayak or coastal walk. Vehicle access is from East Coast Road through Torbay to Long Bay or from SH 1 via Oteha Valley Road. The Okura River can be accessed from Haigh Access and Okura River Roads. The nearest boat ramps are at Torbay, Okura, Wade River, Stillwater and Arkles Bay.

Best time to visit: High tide for swimming, snorkelling and kayaking and low tide for exploring the rocky reefs. Best when there is no wind and little swell on the east coast.

Activities: Snorkelling, scuba-diving, swimming, kayaking, walking.

Facilities: Toilets, changing rooms, cold showers, information kiosk, picnic tables and barbecues are available in the Long Bay Regional Park. The Sir Peter Blake Marine Education and Recreation Centre is near the southern entrance to the regional park.

Rules: Marine reserve rules apply (see page 12). For further information contact DoC's Auckland office on 09 307 9279. Dogs are permitted below the high tide line on Long Bay Beach before 9am and after 7pm during summer and at all times during winter, as far as the northern end of the beach. They are not permitted in the regional park.

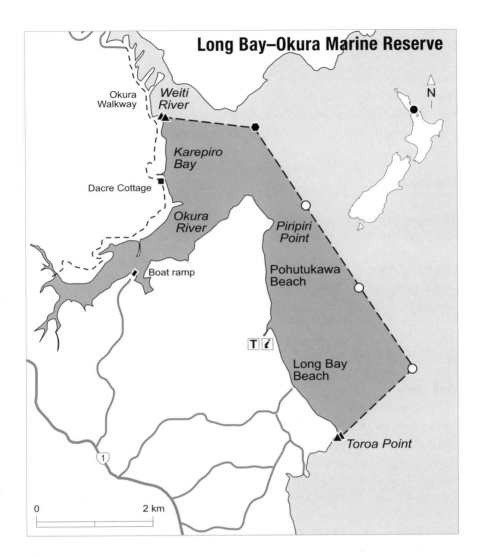

History

Original Maori inhabitants called this area Oneroa: 'long or large area of sand'. Ngati Kahu are the local iwi. In 1862 the Vaughan family purchased some 600 ha of land behind Long Bay. They farmed it for the next 100 years and ran a camping area. In 1965 they sold the land to the Auckland Regional Authority (ARA, now Auckland Regional Council). In 1990 the committee from the Marine Education and Recreation Centre sought protection for the reefs where children snorkelled, and the East Coast Bays Coastal Protection Society was formed. The committee applied for a marine reserve, which was established in 1995.

Topography

The coastline around Long Bay is typical of the inner Hauraki Gulf. It is formed largely of Waitemata sandstones and mudstones and is moderately sheltered. Silt run-off has had a marked effect on the Okura River, shallowing the channels and allowing mangroves to proliferate around the upper reaches.

Marine life

The marine reserve encompasses a variety of habitats, including sandy beaches, rocky reefs, sheltered estuarine mudflats and mangrove forests.

Intertidal zone
To the north and south of the beach are sandstone reefs that shelve to sand at 6 m depth. The coralline algae on the reef is often covered with sediment after heavy rain. Neptune's necklace is the dominant brown seaweed covering large sections of the rock platform from the Okura River right along the reef to Long Bay. Near the high tide mark are barnacles, periwinkles, limpets and chitons.

Rock pools
The rock pools contain a variety of sea snails, including black nerita, cat's eyes, periwinkles, spotted and lined whelks, oysterborers, Cook's turbans and white and dark rock shells. Circular holes in the rocks are created by bivalve rockborers. Sea urchins, usually juveniles, hide under rocks. Hermit crabs are quite common and are the faster-moving shells. Large-clawed half crabs scuttle away when you lift up the rocks. Triplefins and shrimps dart over rock pool bottoms.

Rocky reefs
Encrusting coralline algae covers much of the reef from the mid to beyond the low tide zone. From the low tide mark out to the edge of the sandstone reefs, sponges, sea anemones, bryozoans and orange ascidians cover the shaded areas under the kelp and underhangs. Finger sponges project 30–40 cm above the rock surface at the edge of the reef above smaller golf ball sponges. Spotties are the most noticeable fish, especially in the seaweeds. Snapper and parore are present. Yellow-eyed mullet schools appear around the edges of the reefs in summer. Schools of grey mullet are often seen jumping around the edges of the reefs at the northern end of the reserve. Near the edge of the reefs small crayfish live among sandstone boulders. Seaweeds include stunted common kelp, slender flapjack, green codium, and sea lettuce.

The most abundant seashells are the herbivorous cat's eyes. Predatory lined and spotted whelks hunt over much of the reef. In the sandstone cracks, red rock crabs hide and hermit crabs scuttle busily.

On the sand

Long Bay has recovering populations of tuatua bivalves and sand dollar urchins around the low tide mark. Heart urchins live in the silty sand below the low tide mark. Paddle crabs are common.

Small snapper, goatfish and spotties feed over the seabed and large horse mussel and silky dosinia bivalves live buried in the sand. Near the edge of the reserve, on patches of exposed sandstone, are sponges and stunted kelp. Some small scallops have recolonised below 6 m and goatfish and stingrays feed there. After heavy rain the water is stained and sediment often blankets the seabed.

In the estuary

Near the mouth of the Okura River are large deposits of shell sediments between and above the reefs. Along the northern edge of the estuary sandstone shelves support large numbers of oysters. Filter-feeding tube worms and barnacles live among the oyster shells. Further up the estuary are large expanses of cockle beds, although the shells are small. Several species of whelk shells graze on the beds. Small brown sea anemones attach themselves to loose cockle and pipi shells.

Mangrove trees dominate the upper estuary. Most are small, reaching 2–3 m, and barnacles, oysters and mussels grow on their trunks and roots. Fish species include yellow-eyed mullet, parore and flounder. Often schools of juvenile fish seek the safety of the branches at high tide. Short-finned eels live in the muddy upper estuary. At high tide snapper and goatfish feed along the outer part of the estuary and in summer long and short-tailed stingrays move into the shallows.

Coastal flora and fauna

Pohutukawa trees line the coast and cliffs surrounding the marine reserve. Flaxes and tea tree fill the gaps along the cliff tops but compete with gorse. The Okura River is a feeding ground for white-faced herons, variable and South Island oystercatchers, pied stilts and various wading birds. Kingfishers hunt over

RESEARCH

Numerous studies of the area have been done since the early 1980s. The most recent, by the Leigh Marine Laboratory, looked at 30 sites from Waiwera to Campbells Bay. Sedimentation from run-off was identified as a major problem.

the sand flats and mudflats for crabs and small fish. White-fronted and Caspian terns flutter over the estuary, feeding on schools of small fish that may venture too close to the surface. Introduced rock pigeons nest on the steep cliffs.

Activities

Snorkelling
The reefs at the northern and southern ends of Long Bay are the best spots. Entry is easiest from the sand alongside the reefs as the reef can be slippery.

Scuba-diving
The best diving is at the deeper extremities of the reefs at the northern and southern end of the beach. Entry is best from the sand alongside the reefs or from a boat or kayak. Visibility is limited especially after rain or swells.

Kayaking
Kayaks can be launched off the beach at Long Bay or from Okura, Stillwater or Arkles Bay. The Okura River is best explored on a rising tide. There are various points of interest including the sandspit on the north side of the Okura River and historic Dacre Cottage, built in the 1850s. The beaches between Long Bay and Piripiri Point are easy landing points.

Walking
Coastal Walk (2 hrs return)
The coastal walk travels north along the cliff to Okura River passing Granny's Bay (15 mins), Pohutukawa Bay (30 mins), Piripiri Point (45 mins), and Okura River (1 hr). Lining the tracks are shells from Maori middens. From mid to low tide you can walk around the rocky reefs along the coast. The rocks are sharp and slippery and the cliffs above are unstable, so beware of falling rocks.

Okura Walkway (3 hrs return)
The walkway through the Okura Estuary Scenic Reserve is accessed from the end of Haigh Access or Duck Creek Road. From Haigh Access Road, it follows the Okura River, through native bush, past Dacre Cottage and on to Stillwater.

Oneroa Track (1 hr return)
This walk starts beside the Sir Peter Blake Marine Education and Recreation Centre and winds up towards Toroa Point. The track passes a wartime gun emplacement and there are views north over Long Bay and the Hauraki Gulf.

MOTU MANAWA (POLLEN ISLAND) MARINE RESERVE

Surrounded by industrial Auckland and bisected by a motorway, Pollen Island goes undetected by most of the motorists who cross it daily. It is typical of estuarine and mangrove habitats that are affected by urban run-off but are still rich in life, and it is the best example of its kind in the Waitemata Harbour.

Created: 1995

Size: 500 ha

Boundaries: The North-Western Motorway passes through the reserve between the Te Atatu and Waterview on-ramps. At the Whau River motorway bridge, a yellow triangular marker on the bridge marks the western end of the reserve. The northern boundary runs north-north-east to a yellow buoy, then east to a second yellow buoy and south-east to a triangular marker north of the motorway. A further yellow buoy sits halfway between the northern boundary buoy and the motorway marker. All buoys have flashing lights. All of the area south of the motorway, bordered by the suburbs of Rosebank Peninsula and Waterview and with the exception of Oakley Creek, forms the southern section of the reserve. Around Pollen and Traherne Islands, the landward boundary is at MHWS (high water mark on the highest tides).

Getting there: There is no road access. Walking around the reserve itself is discouraged as the ecosystem is sensitive, but there is a useful walkway/cycleway through the reserve; it is accessed from Waterview, Rosebank Road and Te Atatu South, near the motorway exit. Boats and kayaks can be launched at the end of Bridge Avenue, Te Atatu South. Dinghies and kayaks can be launched around high tide only from the end of Walker Road, Point Chevalier.

Best time to visit: One hour before and after high tide is the best tide for boating and kayaking around the reserve. The Te Atatu–Waterview walkway/cycleway is open at any time.

Activities: Kayaking, walking, cycling, boating.

Facilities: None.

Rules: Marine reserve rules apply (see page 12). Contact Department of Conservation's (DoC) Auckland office on 09 307 9279. Dogs are not permitted in the marine reserve.

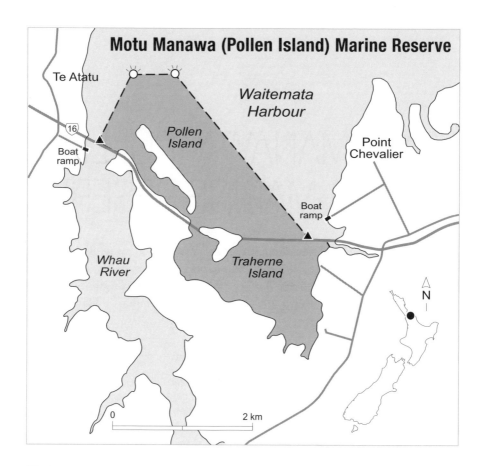

History

Early Maori settlement around the Waitemata Harbour included the Pollen Island area. The head of the Whau River was a portage area for Maori waka (canoes) crossing between the Waitemata and Manukau harbours.

In 1855, a large tract of land at the end of the Whau Peninsula, including Pollen Island, was bought by Dr Denver Pollen, an Auckland medical doctor who became New Zealand Premier in 1875. He started a brickworks in the area – the first of many in the Auckland region. The remnants of a tram or railway line, constructed in the 1920s to carry lime to the limeworks on Rosebank Peninsula, can still be seen. Pollen Island was later the property of the Auckland Harbour Board and was earmarked for port facilities when the Auckland wharfs were outgrown. The advent of container ships meant it was never needed.

In 1990 the Royal Forest and Bird Protection Society proposed the creation of a marine reserve around Pollen Island to protect a representative inner harbour mangrove habitat. They now manage the island under a lease agreement, while the Department of Conservation (DoC) manages the marine reserve.

Topography

The area is geologically significant as it consists of freshwater peat substrate lying beneath marshlands. This formation was created more than 170,000 years ago in the Pleistocene epoch and is visible at low tide on the seaward side of the island. Gradually banks of shells have built up and have stabilised. Both Pollen and Traherne Islands are surrounded by salt marshes and shell banks, mangrove swamps, and tidal mudflats and channels.

Marine life

Salt marshes and shell banks

The north and west of Pollen Island represent probably the best examples of intertidal salt marsh and mangrove habitats in the harbour. Flaxes and cabbage trees sit on solid ground above the salt marsh plants on the muddy flats. Holes in the mud are the homes of thousands of mud crabs, often feeding near the mouths of their burrows. Stunted mangrove plants, few over a metre high, dot this area and slightly larger specimens sit next to wet channels that bisect the flats. Tracks of snails criss-cross the area, the most common being mud snails.

Shell banks have built up and created dry areas that are mainly composed of cockles, although other bivalves like pipi are present. Much of the area to the south of the motorway is only wet on the highest of tides with small mangroves mixed with sea rushes. Eventually this area will dry out completely and glassworts, jointed rushes and raupo will dominate.

Mangrove forest

The higher tidal areas, especially those alongside the channels and waterways, are dominated by mangrove trees and their protruding root systems. The thick mud just below the surface is black and lacks oxygen. To compensate for this mangroves take oxygen from the air with aerial root systems known as pneumatophores. Mangroves live only between the tides, trapping sediment washed down from the surrounding land and causing the seabed to rise and constrict the tidal flow. The trunks and aerial roots have a covering of barnacles and a few oysters. Top shells graze on the leaves and trunks and crabs clamber over them.

Tidal mudflats

Mud snails are very common, leaving visible trails as they digest micro-organisms on the surface mud once the tide has receded. Stalk-eyed mud crabs and common mud crabs move about on the soft mud, feeding on particles and vanish into their burrows if disturbed. Several seashells are present, including mudflat top shells, mudflat horn shells and mud whelks, all of which leave distinct trails across the mud. Mud whelks form large feeding aggregations over anything that has died, homing in using their sensory proboscis which appear from under their shells like periscopes. Fish bodies can be stripped to bare bones over a single tide.

At the half tide and lower bivalve beds start, with cockles the most prolific. Pipi, wedge and harbour trough shells are the other large bivalves. Wedge shells create strange bird footprint-like marks across the mud, with their feeding tubes. The most abundant bivalves are tiny nut shells usually less than 5 mm across that live below the surface. Small harbour limpets live inside large empty bivalve shells, grazing on the algal growth that forms on them.

Tidal channels

On the rising tide, yellow-eyed mullet follow the water in and feed and shelter under the mangroves. They feed on planktonic particles suspended in the water. Sand flounder also move in and sit in the extreme shallows, their camouflage making them almost undetectable until they move.

Schools of parore move under the mangroves on the high tide to feed on the algae around the trees. Some pipi and harbour trough shells live below the low tide where stingrays feed on them. Snapper were once common in the tidal areas but only rare individuals venture into this area today.

Coastal flora and fauna

The intertidal mudflats are important feeding grounds for wading and non-wading birds. Migratory birds such as godwits, knots and sandpipers congregate here before heading to the Northern Hemisphere for winter. During late summer, autumn and winter oystercatchers and wrybills gather before heading to the South Island in spring, where they breed.

The shell banks and salt marshes behind the mangroves create shelter and nest sites for other birds including white-faced herons, pukeko, spotless crake, banded rail, kingfisher, fernbirds and New Zealand dotterels. White-fronted

RESEARCH

Several species lists have been compiled, the most recent by DoC in 2004.

terns are usually in small flocks and you may see a few larger Caspian terns.

Small shrubs and grasses grow on the shell banks and create homes for many of the birds. Wire vine, tussocks and ribbonwoods grow amongst the flaxes, tree daisies and wattles.

Activities

Kayaking

Kayaks can be launched from the public reserve at Walker Road, Point Chevalier and Bridge Avenue, Te Atatu South. The shallow sandbanks to the north of Pollen Island are best paddled at high tide. The most suitable landings are on the shell banks but intrusion into the fragile areas behind the sandbanks is not advisable. These areas are the nesting places of many birds and the salt marsh environment is easily damaged. Allow 2 hours return from either Te Atatu or Point Chevalier starting an hour before high tide.

Walking

Walking through the reserve is discouraged as the ecosystem is extremely fragile. The mud, often knee-deep or deeper, is not pleasant to walk through. The cycle/walkway between Waterview and Te Atatu is a good 1-hour walk each way via Rosebank Peninsula with good views of the marine reserve. A walkway along the eastern side of the Te Atatu Peninsula, accessed from Te Atatu Road, opposite Gloria Avenue or Harbour View Road, overlooks the Whau River and the northwestern part of the reserve.

TE MATUKU MARINE RESERVE

The reserve is an excellent example of salt marsh, sand spit and tidal flats on the largest and most undisturbed estuary on Waiheke Island. Mangrove trees lining the estuary reach 8 m high – a record for the Auckland area – and the tide line is a rich feeding ground for wading birds.

Created: 2005

Size: 690 ha

Boundaries: The eastern boundary extends from the eastern headland of Otakawhe Bay to Kauri Point on Ponui Island. The boundary then extends directly west, then north-east to the western headland of Whites Bay. An oyster farm on the western side of Te Matuku Bay is excluded. The boundaries are marked with yellow triangular markers.

Getting there: Access is by boat or kayak from the mainland. Regular passenger ferries depart from downtown Auckland (45 mins) and Devonport (30 mins) to Matiatia. Car and passenger ferries depart from Half Moon Bay to Kennedy Point. From Matiatia it's a 20-minute drive to Otakawhe Bay and from Kennedy Point it's around 15 minutes. The nearest boat ramp is Kennedy Point on Waiheke Island. If you are setting out from east Auckland there are boat ramps at Beachlands, Omana and Maraetai.

Best time to visit: Low tide is the best time to walk around the reserve and high tide is best for boating and kayaking. Winds from the south and south-west stir up the estuary.

Activities: Snorkelling, scuba-diving, swimming, walking, kayaking.

Facilities: There are no public facilities in the reserve. An outdoor education centre at Otakawhe Bay can accommodate groups of up to 25.

Rules: Marine reserve regulations apply (see page 12). Contact Department of Conservation's (DoC) Auckland office on 09 307 9279. Dogs are not allowed on the shell spits.

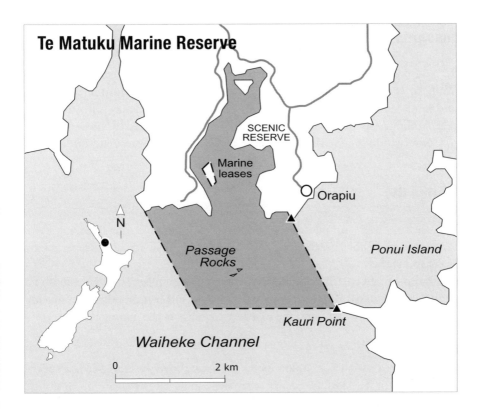

History

Matuku is the Maori name for the bittern, a bird that lives around estuaries. Shell middens show that Maori settled around Te Matuku Bay or at least visited seasonally. Shellfish would have been a major part of their diet. A pa site can be seen on the western headland of Otakawhe Bay.

The island's first European settlers logged the surrounding area for rimu, totara and kauri. Kauri gum can still be found in the shallows of Te Matuku Bay. After the forest was cleared, the land was farmed, but since the mid 1970s the bush has regenerated.

In 1988 a survey identified Te Matuku Bay and surrounding forest as being ecologically significant. In 1991, the Royal Forest and Bird Protection Society and the Waiheke Underwater Club surveyed the waters and suggested three areas to be considered for marine reserve status. Te Matuku Bay was the most widely supported. In 1994 Forest and Bird consulted with various interested parties, including local iwi (tribe) Ngati Paoa and landowners. Their report confirmed widespread support for a marine reserve at Te Matuku Bay, and despite some opposition from bach owners, the reserve was finally created in 2005.

Topography

Te Matuku Bay is the largest and most undisturbed of several estuaries on Waiheke Island. The surrounding reefs and cliffs are a mixture of greywacke and sandstone. The sandspits on the eastern side of the bay are surrounded by regenerating native forest with several bush reserves, managed by the Department of Conservation (DoC), the Auckland City Council and Forest and Bird.

Marine life

The reserve covers a multitude of habitats, including salt marsh, mangroves, sandy beach, pebble beach, rocky shore, intertidal estuary and deeper sandy areas and reefs.

Silt run-off affected Te Matuku Bay when the kauri forests on Waiheke Island were felled around 150 years ago. Further silt run-off from subdivisions around the Waitemata Harbour continue to affect visibility in the marine reserve.

Salt marsh
A healthy area of salt marsh dominates the area above the high tide line on the eastern side of Te Matuku Bay. Several shell spits, largely made up of cockle shells, provide a natural barrier between the salt marsh and the sea, although small channels allow the high tide to penetrate the area. Mud snails and mud crabs are the most common animals on this area of soft mud.

Mangroves
The headwaters of Te Matuku Bay are dominated by mangroves, up to 8 m high, larger than most found in the Auckland area. Mangrove trunks and aerial roots support large numbers of little black mussels, Pacific oysters and acorn barnacles. Grazing cat's eye snails crawl on the mangrove trunks and over the clumps of oysters.

Soft mud, intersected by meandering channels, covers the higher tidal shore

---

RESEARCH

In 1996 a survey by the Marine Department of the Auckland War Memorial Museum found seven species of chiton, 52 snails, 38 bivalves, 34 crustacea, four barnacles, 19 decapods, 10 echinoderms, 21 polychaetes and 31 other animals and plants in the mudflat zone.

Local schools regularly survey the marine life at Te Matuku Bay.

where mud snails are the most common. The thousands of holes in the mud are made by mud crabs, which move out from their burrows but return rapidly if threatened. Kingfishers, which use the mangrove trees as vantage points, are their main predators.

Neptune's necklace seaweed forms loose clumps along the edges of some channels. On the rising tide schools of yellow-eyed mullet move in to feed, and flounder are common close to shore.

Sand flats

The middle to outer bays of Te Matuku have a covering of reasonably firm sand and empty cockle shells. Towards the lower intertidal area are dense cockle beds and patches of eel grass. Wedge shells, another bivalve, are distinguished by the marks resembling bird footprints that are left in the sand by their feeding activities.

The occasional green-lipped mussel attaches itself to larger empty oyster shells. Cushion sea stars are abundant, and the larger spiny sea star is also fairly common. Sand gobies – small estuarine fish – sit almost undetectable until they move among the empty bivalve shells.

On the firmer sand to the south of the bay are horn shells and mud and spotted whelks. Pipi, large trough shells, mud dosinias and small nut shells live in the sand, along with several other bivalve species. Marine worms that live in the sand are predated by flounder that move in on the rising tide.

Rocky tidal reefs and shoreline

The rocky reefs in the bay have good numbers of barnacles, little black mussels, cat's eye snails and grazing periwinkles. Pacific oysters colonise around the mid-tide mark, mixing with tube worms attached to the rocks. Oysterborer snails form dense clusters. Small numbers of chitons live in shaded areas and cracks.

Outer bay and reefs

The sea floor in the outer bay is a mix of sand, gravel and mud. A strong tidal current sweeps between the bay entrance and Passage Rocks. Schools of kahawai are not uncommon and small snapper feed here. Extensive beds of horse mussels live buried in the sand, along with a few scallops. Larger sea snails include Arabic volutes, northern siphon whelks, ostrich foot and turret shells. Several species of marine worm, sea stars and sand dollars also live on the muddy bottom.

Near Passage Rocks, and on the outer reefs, patches of flapjack and common kelp grow down to around 6 m. Some large green-lipped mussels cling to the rocks around the low tide mark. Spotties are the most common fish around these beds, although several species of triplefin live on the rocks below. Sponges, often covered with a layer of silt, manage to exist on the reefs below the seaweeds. A few small crayfish find homes in cracks. There are several deep holes to 25 m and patches of rock support anemones, sponges and grazing nudibranchs.

Coastal flora and fauna

At the northern end of Te Matuku Bay sedges, eel grass and glassworts cover large areas. The salt marsh changes to salt meadow plants and at the most landward end raupo swamp areas mix with mangroves surrounded by grassland. The salt marsh and bushy areas support bitterns, fernbirds, kingfishers and white-faced herons. Fantails and welcome swallows hunt for insects on the shell banks and over the salt marshes.

On the eastern side, towards the mouth of the bay, two raised shell spits are key nesting sites for birds. Waders and seagulls feed over the sand flats. In winter there are pied stilts, South Island and variable oystercatchers and wrybills. In summer godwits, knots and sandpipers feed before returning to the sub-Arctic.

Activities

Snorkelling

Although there is limited visibility, some snorkelling is possible around Otakawhe Bay and by boat around Passage Rocks. Best visibility is on the high tide. Be aware of strong currents around Passage Rocks and stay near the rocks as this area has high boat traffic.

Scuba-diving

The area around Passage Rocks is diveable in spite of the limited visibility. Be aware of strong currents and boat traffic. Best visibility is during high tide.

Walking

Otakawhe Bay (1–1.5 hrs return)

A track on the western side of Otakawhe Bay climbs to a former pa site, then through native bush down to Pearl Bay. At low tide you can walk around the rocky shore to the shell spit. Past the next point the mangrove forests begin.

Kayaking

You can launch at Otakawhe Bay and paddle alongside the shell spit and among the mangrove forest in Te Matuku Bay. From Otakawhe Bay you can also paddle to Ponui Island and around Passage Rock. In southwesterlies, and when the wind opposes the tide, the short, steep waves can be unpleasant.

TE WHANGANUI-A-HEI (CATHEDRAL COVE) MARINE RESERVE

Spectacular rock formations, including Cathedral Cove, dominate Hahei's coastline. Gemstone Bay's unique snorkel trail, marked by a series of buoys, is a great introduction to the reserve's underwater life.

Created: 1993

Size: 840 ha

Boundaries: Large yellow posts mark the boundaries. The western edge extends from Cooks Bluff to Mussel Rock and out to Motukorure Island. The eastern boundary extends from the western end of Hahei Beach to the northwest corner of Mahurangi Island and to the northeast point of the island. The boundary is approximately 300 m out from the post and encompasses Waikaranga Island back to Motukorure Island.

Getting there: From SH 2, turn off onto SH 25, then take SH 25A via Kopu-Hikuai through Tairua, and turn off at Whenuakite to Hahei, following the signs to Cathedral Cove car park. Access to the reserve itself is by boat, kayak or on foot. The nearest boat ramps are at Cooks Beach and Whitianga. Boats can be launched using tractors at the southern end of Hahei Beach.

Best time to visit: Late summer when the water is clearest but it can be good any time of the year during settled weather. Strong easterly winds, swells and heavy rain limit underwater visibility. Visit Cathedral Cove at low tide, but swimming and diving can be done on any tide.

Activities: Snorkelling, scuba-diving, swimming, kayaking, walking.

Facilities: Snorkel trail, toilets, changing rooms, information kiosk, coastal lookout, snorkel and dive gear hire.

Rules: Marine reserve regulations apply (see page 12). Contact Department of Conservation's (DoC) Thames office on 07 867 9180 for more information.

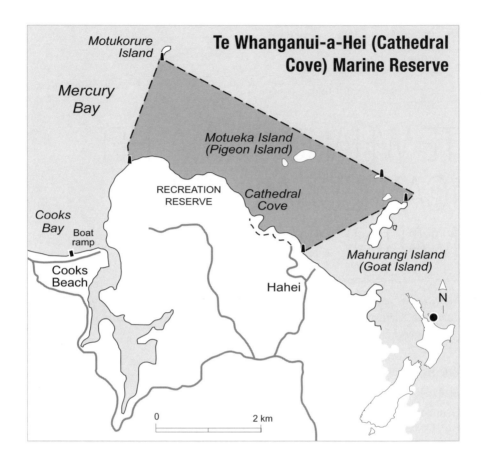

History

Hahei is part of the Mercury Bay area settled by Hei, a teacher or skilled person who arrived on the Te Arawa waka (canoe) around 1350AD. Hei's descendants, Ngati Hei, retain strong ancestral and spiritual attachments to the reserve and are guardians of the area.

The explorer James Cook sailed into Mercury Bay in 1769 and made contact with local Maori. From 1862 the area was farmed by European settlers.

The idea of a marine reserve for Coromandel was first raised in the early 1980s when DoC sent out a questionnaire on the placement of a marine reserve. Feedback from the Coromandel favoured the Mercury Bay area and the variety of habitats around Hahei suggested it was the best option. The 1989 proposal was supported by Ngati Hei and most local residents. Because of opposition by commercial fishers and a few Hahei residents, the beach was not included in the marine reserve, which was gazetted in 1993.

Topography

The coast is a mixture of glassy volcanic ash and pumice known as ignimbrite, which erupted from long-extinct rhyolitic volcanoes. Over the years the cliffs have eroded creating a spectacular coastline. There are many caves and archways, the most well known being Cathedral Cove.

The range of underwater habitats includes rocky reefs, soft sediments, caves and archways. The islands of Motukorure, Moturoa, Motueka and Mahurangi, all of which lie within the reserve, are diverse and well worth exploring. There is usually somewhere for boats to shelter.

Marine life

The snorkel trail at Gemstone Bay gives the observer a good picture of the coastal reefs of the reserve. Similar terrain exists at neighbouring Stingray Bay and on the western side of Mahurangi Island.

Intertidal zone
Large, rounded granite boulders in an array of colours dot the shoreline between small patches of sandy beach. Because of boulder movement in storms, few organisms can secure a permanent residence.

RESEARCH

Monitoring of crayfish numbers began in 1996. In 2003 crayfish numbers were 15 times higher than in comparable areas outside the marine reserve. In fact, crayfish numbers outside the protected area have declined by around 90 per cent since 1996.

Fish numbers have been monitored since 1997 using baited underwater video and underwater visual counts. In 1999 snapper larger than the legal catchable size were 12 times more abundant than in comparable areas outside the marine reserve. Smaller snapper numbers were not noticeably different.

During 1999 and 2000 reef communities were monitored at five sites inside the marine reserve and five sites outside the reserve. Comparisons of seaweeds at depths of 4–6 m showed that seaweeds existed in quantities three to four times higher than outside the reserve boundaries. Sea urchin numbers were no different in 1999 but in 2000 showed a higher incidence of crevice occupancy in the reserve and a lesser number generally, due to the return of their predators, crayfish and snapper.

The tidal zone below the cliffs has sea lettuce and small brown seaweeds with grazing limpets and chitons. Around the offshore islands of the reserve there is similar life, depending on the level of exposure to the Pacific swells. The splash zone in most cases is dominated by several species of barnacle, small seaweeds and clinging bivalve shellfish such as rock oysters and little black mussels. Various small green and red seaweeds dominate the low tidal zone and below.

Shallow bays

Small fishes like triplefins and spotties are common and dart around among the limpets on the boulders. Cushion stars and sea urchins hide in the crevices between boulders and black nerita snails graze on the tops. Paua, both black and yellow foot, live on the sides of and under boulders.

At depths of 4–5 m the boulders give way to shallow broken reef with flapjack seaweed. Red crayfish live in the gaps under the boulders and in the reefs, with some individuals in Gemstone Bay weighing in at several kilograms. A variety of other seaweeds grow on the reefs and grazing sea shells, especially top and Cook's turban shells are present. At depths of around 5 m common kelp begins to grow.

Snapper, from 10 cm striped juveniles to large adults over 5 kg in weight, feed on the reefs and sand. Some will approach divers but most are still inclined to be flighty.

Kelpfish and marblefish are common on the reefs near the kelp and red moki, leatherjackets and goatfish are everywhere.

Eagle rays, some more than a metre across, often rest on the sandy areas, right up to the edge of the shallow boulders. The sandy areas are pockmarked with the digging holes of fish and eagle rays.

Some rocks and pinnacles climb to near the surface and are dominated by sea anemones. Common white anemones are prolific with colourful jewel anemones on the underhangs. Grazing invertebrates, such as the common sea urchin, form large groups, along with turban, top and tiger shells. Patches of green-lipped mussels up to 12 cm nestle almost invisible among sponges and anemones near the reef top.

Kelp clings to the rocks below 5 m and continues down to depths of 20 m. Around Kingfish Rock marblefish congregate in large numbers, usually resting on the rock faces among the anemones and sponges. Their feeding technique, which consists of grasping mouthfuls of seaweed and allowing the surge of the swell to tear the seaweed clear of the rock, is worth observing. They then settle against the rock to eat the torn-off seaweed.

In summer the schools of thousands of mackerel around the reefs and pinnacles are predated by kingfish, which feed as individuals or in schools.

The largest kingfish are well over a metre in length. Schools of blue maomao, sweep and demoiselles mingle with mackerel and koheru in midwater. Large sandy areas dominate between depths of 10 and 15 m but change back to reefs around the offshore islands.

Deeper reefs

The reef walls below the kelp forest are encrusted with anemones, sponges, ascidians, soft corals and hydroids. The latter two species are preyed on by two of New Zealand's most spectacular endemic nudibranchs. Apricot nudibranchs, which grow to nearly 15 cm in length, feed on the soft coral known as dead man's fingers. Hydroids, which extend from the walls like miniature bonsai trees, are grazed by Jason nudibranchs. The Jason nudibranch can grow up to 10 cm in length, and is bright pink with white finger-like growths, or cerata, along the top of its body.

The habitat at the base of the cliffs is mostly large boulders, usually covered with sponges. Yellow moray eels are the most common morays. Other predators, such as dwarf scorpionfish, sit among the invertebrate life, almost invisible until lit by a diver's light or strobe.

Schools of butterfly perch and two-spot demoiselles school above the deeper reefs near the islands. Scarlet, banded and Sandager's wrasse, along with red pigfish, add colour to the deeper realm.

At depths of 25 m sponge gardens begin to dominate, especially in areas unaffected by the Pacific swells.

Marine mammals

A growing number of young male New Zealand fur seals winter over on Mahurangi Island. Midden material excavated from local pa sites shows that seals were once a major part of the Maori diet. Bottlenose dolphins are the only other marine mammal commonly seen in the reserve. Common dolphins occasionally skirt the outer boundaries and pods of orca (killer whales) are rare visitors.

Coastal flora and fauna

Pohutukawa dominate the coastal cliffs and tui are the most common birds, especially when the pohutukawa are in flower during summer. Pied shags nest in the pohutukawa and feed in the waters of the reserve. Australasian gannets are regular visitors and dive and feed in the waters of the reserve. Little blue penguins and several species of petrels and shearwaters also feed in the area and nest on Mahurangi Island.

Activities

Snorkelling
The best snorkelling from shore is at Gemstone Bay and Stingray Bay 5–10 minutes walk from the Cathedral Cove car park. There is good snorkelling around the offshore islands from a boat or kayak and in winter the New Zealand fur seals are great underwater playmates.

Gemstone Bay Snorkel Trail
DoC has put in place a snorkelling trail at Gemstone Bay. Four marker buoys with information panels depicting which species inhabit each area, are anchored from 50 m to 165 m from shore. Each buoy has grab handles and provides good flotation while reading the information. Take care walking over the slippery boulders when entering the water. The boulders near the water are the best place to fit dive fins as walking across the boulders in fins is dangerous.

Scuba-diving
The reserve is not easily accessible from land when you are carrying scuba gear. The easiest diving is from a boat or kayak. There are good dive sites around the offshore islands and several pinnacles, the best being Kingfish Rock.

Kayaking
Kayaks can be launched off Hahei Beach. You can either paddle along the coast to see the caves, archways and rock formations or around the numerous islands that dot the area. Mahurangi Island has a landing area but be aware of the seals that haul ashore in winter. While they're fun to paddle with they should not be approached on land as they can be aggressive.

Walking
Cathedral Cove Track (1 hr return)
The track begins at the Cathedral Cove car park and is well signposted. The track is a 10-minute downhill walk through bush to Gemstone Bay and it's a further 5 minutes through pohutukawa trees to Stingray Bay. Cathedral Cove, with its spectacular rock formations, is 30 minutes from the car park and is best visited at low tide when you can walk through the arch. The track passes a pa site just above Cathedral Cove.

Mahurangi Island
Mahurangi Island, the largest of the islands enclosing the bay, is a recreation reserve and landing is permitted. Various walking tracks criss-cross the island.

MAYOR ISLAND (TUHUA) MARINE RESERVE

Lying well offshore, Mayor Island benefits from the presence of the subtropical East Auckland Current. Many of the marine species seen here are not found on the coast and the water is much clearer. The island's volcanic history, rock formations and underwater gas vents make it unique.

Created: 1992

Size: 1060 ha

Boundaries: The western boundary at Tumutu Point, which extends north-west, and the eastern boundary at Turanganui Point, which extends north-east, are both marked by two triangular yellow markers. When the markers are aligned vertically these are the eastern and western boundaries to a point 1 nautical mile offshore. The northern boundary is 1 nautical mile offshore from the nearest part of Mayor Island.

Getting there: The marine reserve is at the northern end of Mayor Island, approximately 35 km from Tauranga, and is accessible by boat. The nearest boat ramps are at Tauranga and Whangamata.

Best time to visit: Summer and autumn when the water is warmest and clearest.

Activities: Snorkelling, scuba-diving, swimming, kayaking, walking.

Facilities: The island has cabin and backpacker accommodation.

Rules: Marine reserve regulations apply (see page 14). Contact Department of Conservation's (DoC) Tauranga office 07 578 7677 for further information. No fishing is allowed inside the marine reserve but large gamefish, such as billfish, tuna and sharks (not kingfish), which have been hooked outside the reserve can be played and landed if they stray into the reserve. The remaining area surrounding Mayor Island is a restricted fishing zone that extends to 1 nautical mile offshore. Set nets and long lines are prohibited. Contact Ministry of Fisheries 0800 478 537 for further information.

The island is a wildlife refuge and landing is at South East Bay only. Contact the Tuhua Trust Board 07 579 5655 for further information.

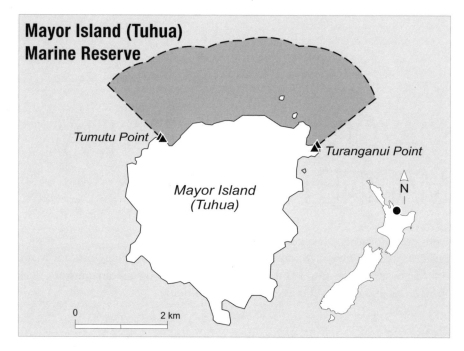

History

Maori oral history tells of Ngahue, a Maori explorer who travelled from Hawaiki to Tuhua. Obsidian – volcanic glass – found on Tuhua was used by Maori for tools. The Maori named the island after the ancient name for the eastern Polynesian island of Me'etia where obsidian was also found. The English name (Mayor Island) was given by explorer James Cook in 1769, when he referred to the island as 'The Mayor'.

The Whanau-a-Tauwhao hapu (subtribe) of Ngaiterangi made the island their home, often having to defend it against invasion. Their pa sites on the high headlands enabled them to do this successfully; however, an epidemic in 1862 decimated the population and by 1916 the island was uninhabited.

The waters surrounding the island became famous in the 1920s for recreational and big game fishing. Later commercial fishing depleted fish stocks around the island.

The island, owned by the Whanau-a-Tauwhao, has been administered by their representatives, the Tuhua Board of Trustees, since 1949. Concerns about the depletion of marine life prompted the board to propose the creation of a marine reserve with a complementary restricted fishing area around the island. These proposals received support and took effect in 1992 after investigations by DoC and the Ministry of Fisheries.

Topography

Mayor Island is roughly circular and 3–4 km in diameter. It is the remains of a volcano that has gradually risen from the sea. A massive eruption occurred 35,000 years ago and volcanic activity continued with another very large eruption around 6300 years ago. From this point activity gradually decreased with only a few hot springs active around the shore today. Inside the large crater are two lakes, one green and one black. Black volcanic glass, known as obsidian, sits in layers among the solidified lava flows. Mayor Island is the only place in the world where the mineral tuhualite is found.

The marine reserve protects representative examples of most habitats around Mayor Island including sand, gravel and boulder flats, boulder reefs, vertical rock faces and deep reefs. Eroded pinnacles, dotted around the island, are the remains of lava dykes.

Marine life

Mayor Island is affected by the subtropical East Auckland Current that brushes northeast New Zealand. The current brings fish and invertebrate species from the subtropics to the reserve.

Shallow rocky reefs

At the marine reserve's western end, a boulder-strewn reef runs away to the northwest below Tumutu Point. Opoupoto Bay, to the east is a clean, sandy bay with projecting boulders covered in red, green and brown seaweeds. The cliffs are composed of spectacular vertical columns of lava surrounding a cave (hence its alternative name of Cathedral Bay).

Dense concentrations of sea urchins graze the zone between 3 m and 6 m depths. Their numbers have not diminished as they have in some other marine reserves, where crayfish and snapper have returned and controlled them.

Forests of brown common kelp cover the reefs under the cliffs and along the coast from about 5 m depth. There are many cracks and crevices under the kelp forest which support anemones, ascidians, hydroids and other encrusting colonial animals. Trumpet shells and several species of tiger shells are also found here. Black angelfish feed on green sea lettuce and leatherjackets, kelpfish, spotties and red moki are common. Goatfish dig in the sand and small schools of trevally, usually juveniles, feed here.

At Orongatea Bay the shallow bottom is mainly rounded boulders and sandy gravel with a covering of seaweeds, including tall brown flapjack seaweeds.

Large black spiny sea urchins mix with common urchins and replace them at below 10 m. On or under the shallow rocks are chitons, cat's eye snails, limpets, brittle stars, sea stars, sea cucumbers and red rock crabs.

Snorkelling is excellent over the shallow rock platforms coated with seaweeds, among schools of sweep and blue maomao. Several hot springs bubble up from the gravel close to shore.

Offshore reefs

The sea floor drops away steeply to around 30 m or more. The reefs are angular rock platforms, mixed with sandy patches and more rounded boulders. Common kelp covers almost all this area. Some steep pinnacles rise up with schools of butterfly perch and pink maomao around them. Tokopapa and Tawekaweka Islands sit offshore from Orongatea Bay and further out Two Fathom Reef comes up to 4 m. Beyond here the reef walls drop steeply to sand at around 50 m. Sponges dominate the reefs below the kelp gardens which diminish below 30 m. Schools of demoiselles hover above the sponge gardens and schools of kahawai and mackerel are common.

Turanganui Bay also known as Elephant Bay, at the eastern end of the reserve is relatively sheltered, especially from southwesterlies. It has steep, surrounding cliffs and bordering reefs. The shallow boulder habitat gives way to coarse gravel and sand at 15 m.

More than 60 species of fish have been recorded in the reserve, including Lord Howe coralfish, sharp-nosed pufferfish and several wrasse more typical of warmer waters. Blue cod and blue moki, more often associated with the cooler waters of southern New Zealand, also make an appearance.

The deep cracks and crevices among the boulders create habitats for crayfish and both red and the rarer green or packhorse crayfish are found here. Moray eels – usually yellow moray, up to 1.5 m long – often haunt similar corners. An encounter with the mosaic moray is a highlight of any dive.

RESEARCH

Since 1993 the Bay of Plenty Polytechnic's Marine Studies Department has conducted a yearly survey of reef fish using an underwater visual census. In 2004 they began studying predatory reef fish using a baited underwater video. Both monitoring programmes show that the marine reserve has experienced a limited recovery of targeted reef fish species, with higher numbers in the reserve than in the recreational fishing area. However, increases in fish numbers do not match those of Te Whanganui-a-Hei and the Poor Knights Islands Marine Reserves, possibly due to illegal fishing.

At depths of 20 m or more, solitary cup corals with anemone-like tentacles and extensive areas of lace corals are found. Black coral colonies have been recorded at the base of cliffs around 50 m or more.

Two-spot demoiselles, porae, scarlet and Sandager's wrasse hang around the reefs and brilliantly coloured splendid perch swim below schools of pink maomao. Long-finned and giant boarfish frequent the sandy areas below 30 m. They use their snouts to uncover worms and crustaceans buried in the sand. Big kingfish, often in schools of a dozen or more, race through the schools of koheru and mackerel in organised feeding forays.

Thanks to their camouflage abilities, octopus lie undetected among the boulders and the related broad squid moves in to lay its sausage-like white eggs on the kelp and rocks. The paper nautilus looks like an octopus but creates white shells as egg cases. Although not as common as they once were, these molluscs (up to 30 cm in length) move in around November. Large numbers of their white shells are left on the seabed after the animals have laid their eggs and died.

Marine mammals

Being so far offshore the island has regular visits from dolphins. Large grey bottlenose dolphins are usually seen in pods of around a dozen. Common dolphins, up to 2 m long, and with their distinctive grey-and-cream bodies, form pods of up to 1000 animals. Orca (killer whales) are less frequent visitors and Bryde's whales are often seen. Migratory whales, travelling from the Antarctic to the tropics to breed and mate, pass the Bay of Plenty coast and close to Mayor Island. Blue, fin, sei, humpback and minke whales have all been recorded.

Coastal flora and fauna

Mayor Island has been free of introduced predators since 2002 after DoC and the Tuhua Board of Trustees began an eradication programme in 2000. Tall, spindly pohutukawa cling precariously to the cliffs. Rewarewa and mahoe mix with manuka, rangiora and tree ferns. Pied shags nest on the pohutukawa and feed around the reefs. Shearwaters and petrels nest on the island and form dense feeding flocks in the waters. Often Australasian gannets from White Island feed in the masses of small fish that school near the surface just offshore.

Activities

Snorkelling

The best sites are in Orongatea Bay, Turanganui Bay and Opoupoto Bay. Access is by boat only; there is usually a sheltered anchorage somewhere in the reserve.

Scuba-diving

There are dive sites to suit all skill levels from the shallow reefs close to Tuhua to the deeper exposed Two Fathom Reef which drops steeply to beyond diving depths where anything could swim past.

Kayaking

Because of the distance offshore, many kayakers transport their kayaks by boat to explore the island. The island's circumference is around 15 km and, if weather conditions permit, is easy to kayak. There are many rock stacks to kayak around, but landing is restricted to South East Bay.

Walking

No permit is required to visit the island, which is privately owned and a wildlife refuge.

Island circuit (full day)

The track begins and ends at South East Bay with two options through the forest to the crater rim. Each is a steep climb before levelling out and following the rim of the crater with some good views into the crater and over the sea. The track descends into the crater, which is swampy in places and passes the black and green lakes, Te Paritu and Aroarotamahine. There are a number of smaller steep tracks down to Akakura Bay, Otura Bay and Crater Bay. Tramping boots are essential and take plenty of water.

Above: *Tawharanui Marine Park protects a mix of sandy beach and rocky reefs next to a regional park.*

Right: *Jason nudibranchs, found right around the New Zealand coast, usually feed and lay their eggs on tree-like hydroid colonies, but are unaffected by the hydroids' stinging cells.*

Right: *Tidally flooded mangroves are important coastal nurseries for young fish and other marine life.*

Below: *John Dory blend in with seaweeds, enabling them to get closer to the smaller fish on which they prey.*

Bottom left: *Hairy sea hare at Long Bay. In some summers these molluscs proliferate on the northern coasts.*

Bottom right: *The marine reserve at Long Bay–Okura encloses typical Auckland North Shore habitats.*

Top: *The shell spit is a favourite nesting place for birds in Te Matuku Marine Reserve, Waiheke Island.*

Above: *A tiger shell nestled among anemones at Te Whanganui-a-Hei.*

Right: *Yellow-eyed mullet school around mangroves as they feed in estuaries at Waiheke Island.*

Above left: *Red rock crabs inhabit cracks in intertidal reefs, emerging at dusk to feed.*

Below left: *Below the kelp zone, sponges in all colours, shapes and sizes are the dominant organisms.*

Top: *Sharp-nosed pufferfish are typical of subtropical migrants brought to the North Island's east coast by the East Auckland Current.*

Above: *New Zealand's most colourful sea star is the firebrick, another subtropical migrant.*

Above: *Mosaic moray eels are the most spectacular of five moray species found around the northern coast of New Zealand.*

Left: *The northern scorpionfish's mottled patterns provide the perfect camouflage.*

SUGAR LOAF ISLANDS (NGA MOTU) MARINE PROTECTED AREA

The large numbers of seabirds on the Sugar Loaf Islands make them the most important nesting site in the Taranaki region. New Zealand fur seals also breed on the islands and dolphins and whales are frequent visitors. Underwater there is an equally interesting mix of South Island and North Island species.

Created: 1991
Size: 749 ha
Boundaries: From south of the Herekawa stream mouth on Back Beach to north of Paritutu (excluding most of Back Beach north of the Herekawa Stream). Outer extension is 900 m seaward from Waikaranga (Seal Rocks) and Motumahanga (Saddleback Island).
Getting there: From New Plymouth follow St Aubyn St to Breakwater Rd. Turn left into Ngamotu Rd then right into Centennial Drive. This takes you to Paritutu Centennial Park, which overlooks the islands and rocks. The reserve is accessible on foot, by boat or kayak. The nearest boat ramp is near the lee breakwater inside Port Taranaki.
Activities: Snorkelling, scuba-diving, kayaking, limited recreational fishing, wildlife tours.
Best time to visit: Late summer is the best diving time when clearer water moves down from the north. Best during easterly winds when there is no swell.
Facilities: None.
Rules: No landing on the islands or rocks without a DoC permit, contact the New Plymouth office 06 759 0350. Some fishing is allowed, but the use of nets, set lines or lines with more than three hooks is prohibited. In the conservation area that extends 500 m off Seal Rocks, fishing is restricted to trolling or spear-fishing for kingfish and kahawai and no collecting of marine life is permitted. Elsewhere in the marine protected area, commercial fishing is restricted to trolling or spear-fishing for kingfish and kahawai. Normal recreational daily bag and size limits apply. Contact Ministry of Fisheries 0800 478 537.

Sugar Loaf Islands (Nga Motu) Marine Protected Area

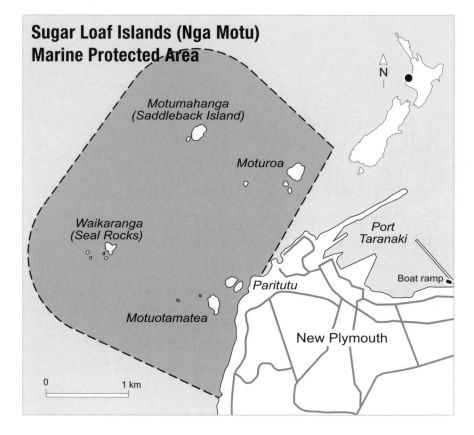

History

The islands, with the exception of Pararaki, were once occupied by the Te Atiawa and Taranaki tribes. Many battles took place here as the area had strategic significance to Maori; the islands were a valuable source of seafood.

Explorer James Cook named the islands in 1770 after their resemblance to the cone-shaped 'loaves' in which stored sugar was heaped in Europe. A few European whalers and traders lived around the settlement of Nga Motu (after which the marine protected area is named), but left after a battle between local Te Atiawa and invading Waikato tribes. Whaling began again in 1841. Port Taranaki was built in 1881 and Moturoa was quarried for material used in the breakwater.

The area became the Sugar Loaf Islands Marine Park in 1986 and the Sugar Loaf Islands Marine Protected Area (SLIMPA) – established by the Parliamentary Act of the same name – in 1991. The purpose of the Act was to protect the islands and the marine habitats while still allowing recreational activities. The islands, rocks and waters are managed by DoC and fishing is managed by the Ministry of Fisheries.

A proposal to change the southern part of SLIMPA into a marine reserve and to extend the boundary southwards to the Tapuae Stream is currently under consideration.

Topography

The geology of the islands marks the earliest phase of Taranaki's volcanic activity around 1.75 million years ago. All reefs and islands are eroded andesite domes. A mineral associated with guano, taranakite, was discovered on the Sugar Loaf Islands. Underwater habitats include rocky reefs with caves and crevices, pinnacles, boulders and sand areas.

Marine life

The marine life is influenced by the warm West Auckland Current which moves down the west coast from just north of New Zealand. There is an interesting mix of warm water species and southern water species. The islands provide some shelter from the prevailing westerly winds and swells.

Shallow reefs

Close to shore, common kelp and long flapjack seaweeds thrive but these drop away noticeably offshore. Much of the habitat close to shore is affected by silt run-off, which reduces visibility for divers, especially after heavy rain and large swells. Mussels live close to the tops of the reefs among gardens of common and jewel sea anemones. Some paua are found on the reef walls and the sides of boulders but they are not common. Underneath the kelp forest the dominant plant life is pink coralline algae. Spotties and kelpfish are common and larger, similarly marked marblefish are seen regularly.

Sea urchins have created bare areas of rock, known as urchin barrens, by

RESEARCH

Some species lists and habitat maps have been produced for the area. The most recent maps were produced by the University of Waikato and DoC from side scan sonar readings, showing the contours, depths and habitats of the marine protected area.

overgrazing the kelp forest. Encrusting invertebrates are dominant on the rock walls between 5 and 30 m. Typical are hydroids, anemones and sponges, with 33 species of the latter having been recorded. Clown nudibranchs are the most obvious sea slug but several other species are also common in this area.

The cracks and crevices in the rocks house some large red crayfish, especially in the conservation area around Seal Rocks and Tokatapu (Hapuka Reef).

More than 80 species of fish have been recorded in the marine protected area. In summer schooling koheru, kahawai and mackerel are hunted by kingfish.

Resident schools of blue maomao and demoiselles mix with silver drummer, snapper and trevally. Copper moki, more common south of this area, school with blue moki. Black-and-white striped magpie perch, better known from Australia, are occasional visitors.

Deeper reefs

On the deeper reefs, sponge gardens dominate with finger, golfball and encrusting sponges. Patches of sand intersect the rocky reefs and goatfish forage with their finger-like barbels. Gurnard, tarakihi, blue cod, scarlet and banded wrasse are the most common fish.

The deeper areas around the outer islands and rocks have less silt and more encrusting life. Colonial bryozoans, ascidians and hydroids live among the sponges. On the branches of white, bonsai-like hydroids, endemic sea slugs – Jason nudibranchs – are found, along with their matching pink rosettes of eggs. Dark ancorina sponges have patches of bright yellow zoanthids, which look like sea anemones, growing on them. Large tiger shells live on the sponges and in the cracks in the rocks. Sea perch and dwarf scorpionfish rest amongst the encrusting life and small grouper-related red-banded perch share this habitat.

Marine mammals

New Zealand fur seals haul out and breed on the islands and rocks, with over 400 animals present during winter. It's not unusual when diving or kayaking to have one swirling past. Common, bottlenose and the rare Maui dolphins are known around the islands, this being the southernmost range for the latter species. Humpback whales, which migrate through Cook Strait, move close to the islands and southern right whale sightings have increased in recent years. The top mammalian predator, the orca (killer whale), follows the west coast and large pods of pilot whales have been observed moving past.

Coastal flora and fauna

Nineteen species of seabirds occur in the area, making it the most important seabird locality on the Taranaki coast. Breeding species include sooty, flesh-footed and fluttering shearwaters, red-billed gulls, white-faced storm, diving and grey-faced petrels. White-fronted terns, black-backed gulls, little blue penguins and black shags nest on some islands and rare reef herons are regularly seen on the outer islands that are rat-free.

In spite of their exposed situation the islands play host to a wide variety of plant life. The most common are flaxes and taupapa scrub, with some karo and cabbage trees. Pohutukawa grow on the southern and eastern slopes of the islands. Cook's scurvy grass, almost extinct on mainland New Zealand, lives on Moturoa and Motumahanga Islands.

Activities

Snorkelling
Because of the exposed nature of this coast, snorkelling from the beach is rarely possible. From a boat the reefs around the larger offshore islands are good value with school fish and fur seal encounters adding to the excitement.

Scuba-diving
This is only possible from a boat and the best diving is in the conservation zone around Waikaranga (Seal) Rocks where crayfish numbers have increased markedly. In summer when the water is clear there are large schools of fish.

Kayaking
Launching is best at Nga Motu Beach, in the Port Taranaki breakwaters. Landing is not permitted on any of the islands or rocks. Onshore winds and large swells make many days unkayakable. Always check weather conditions beforehand.

Walking
Beach Walk (30 mins return)
From the Centennial Park it's a steep descent to the beach, then an interesting tide line ramble along Back Beach to Herekawa Stream.

Paritutu Centennial Park
There are several short walkways to lookout points with good views over the islands. Seals can often be seen on the rocks below and the abundant bird life is often feeding above the fish schools, not far from shore.

TE TAPUWAE O RONGOKAKO MARINE RESERVE

This reserve was created in a joint venture between local Ngati Konohi and the Department of Conservation (DoC). Local schools use the reserve for marine studies and are monitoring the steadily increasing marine life.

Created: 1999

Size: 2452 ha

Boundaries: The northern boundary is south of the Waiomoko River. Two yellow triangular markers, when aligned vertically, mark the boundary, which extends 2.7 nautical miles out to sea. The southern boundary is south of the Pouawa river mouth. This boundary is similarly marked by two yellow triangular markers and extends 1.9 nautical miles out to sea. The line between the two seaward points is the seaward boundary.

Getting there: From Gisborne take SH 35 to Pouawa, about 16 km north. To access the reserve turn right down the beach track just past Pouawa. There is no public road access to the northern end of the marine reserve. The reserve itself is accessible by boat, kayak or on foot. The nearest boat ramp is at Tatapouri, 5 km south of the reserve. Kayaks can be launched from the beach at Pouawa.

Best time to visit: When the wind is offshore and there is no swell. The water is clearest after a period of calm or light offshore winds with no rain. Very low tides are the best time to explore the intertidal reefs.

Activities: Snorkelling, scuba-diving, swimming, beach walking, kayaking.

Facilities: None.

Rules: Marine reserve regulations apply (see page 12). Contact DoC's Gisborne office 06 867 8531 for further information.

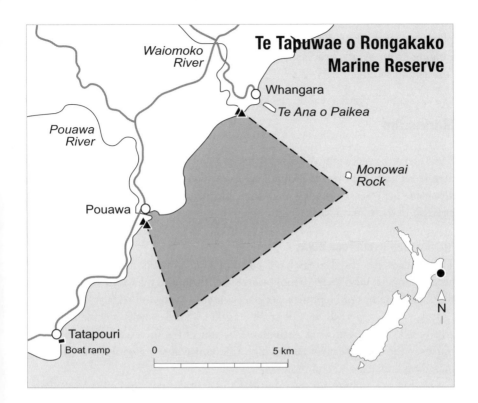

History

Te Tapuwae o Rongokako means 'the footprint of Rongokako'. According to Maori oral history, Rongokako was a giant man who could stride great distances. Several depressions on the sandstone reefs between Hicks Bay and the Mahia Peninsula, including one south of Whangara are, according to legend, the footprints of Rongokako.

The marine reserve resulted from a joint application by the local iwi (Ngati Konohi) and DoC, and was the outcome of nine years of hard work and outstanding leadership by the late Jack Haapu, a kaumatua (elder) of Ngati Konohi. It was Jack's wish to see his tribe's traditional food-gathering area protected 'as a nursery and a sanctuary for the benefit of future generations'.

Topography

The area covered by the reserve is regarded as one of the most spectacular and interesting marine environments in the Gisborne region. Sandstone reef platforms, typical of the East Coast, hold a rich diversity of marine habitats, ranging from sandy beaches to intertidal reef platforms, inshore reefs, kelp forests and sediment flats.

Marine life

The reserve covers eight different types of habitat and is typical of the marine area from East Cape to the Mahia Peninsula. The East Cape Current runs south alongside the reserve bringing planktonic life including large numbers of juvenile crayfish in their larval form.

Shallow reefs and rock pools

The shallow intertidal areas have barnacles and limpets in profusion. A few golden limpets, with their 50 mm orange shells, cling to the rocks at the mid tide mark and patches of Neptune's necklace seaweed surround rock pools.

Lift any rock and dozens of crabs scuttle away. Seashells include cat's eyes and whelks. Faster-moving shells have hermit crabs in residence. Snorkelling out over the shallow reefs on the high tide you will see a variety of seaweeds, including flapjack and common kelp.

Grazing sea urchins dot the intertidal and subtidal flats and rock pools. Paua live on the reef underhangs and under larger rocks. Large black slugs known as shield shells, up to 15 cm long, crawl across the rocks. Their white shells, about a third the length of the slugs, are hidden inside skin folds of the slug and are only uncovered if touched.

RESEARCH

Crayfish have been tagged inside or near the marine reserve as part of a long-term project initiated in 2003 to study their movement patterns and growth rates. The tagged crayfish are monitored regularly by DoC and local fishers. Male crayfish move inshore between summer and winter, then offshore between winter and summer. After moving offshore, many crayfish have returned to within 100 m of where they were tagged the previous year. Initial data suggests growth rates of males slows as their size increases. Female growth rates were low over the same period, perhaps due to them bearing eggs over the survey time.

In 2005, a fish count using baited underwater video had problems as large numbers of crayfish swarmed around the bait during daylight, making fish counts impossible by this method.

Southeast swells from storms in the Southern Ocean wash huge amounts of kelp, sponges and other marine life onto the beach. Rare visitors in late spring are pure white paper nautilus shells, up to 20 cm long, the egg cases of an octopus-like creature, which moves into the shallows to breed and then dies.

The water is often cloudy due to sediment in suspension from the sandstone reefs which is then carried into the reserve by the contributing rivers. Often large rock pools from the mid tide area provide clearer water and a good selection of marine life. One of the best pools is a large shallow area near the southern end of the reserve, which is deep enough to snorkel in.

Marblefish, red moki, spotties, kelpfish, banded wrasse, butterfish and parore feed around the shallow seaweed areas. Some of these can be seen in the larger rock pools, along with small to medium-sized crayfish. Very small red crayfish, between 5 and 10 cm in length, crowd into these shallow reefs during their seasonal migrations.

Subtidal reefs

The deeper reefs, between 10 and 20 m, are home to large numbers of red crayfish which have increased considerably since the reserve was created. Dense forests of common kelp cover some of these reefs but generally give way to sponge gardens at around 15 m. Among the kelp forests are banded and scarlet wrasse and butterfish. Schools of sweep mill about above the kelp and leatherjackets nip at the encrusting life on the rocks.

Dwarf scorpionfish and sea perch sit camouflaged among kelp holdfasts and sponges. Sponge gardens are shared with other colonial invertebrate life such as hydroids, anemones, ascidians and soft corals. Below 20 m schools of tarakihi begin to appear and butterfly perch hover over and around the reefs. Larger blue cod, blue moki and gurnard live in these deeper areas.

The reefs gradually give way to muddy gravel seabeds, with some reefs in the outer areas of the reserve. Crayfish are quite common in any available crack or crevice in the deep reefs and often wander across the sea floor in daylight.

Marine mammals

New Zealand fur seals are occasional visitors in winter although there are no breeding colonies in the area. Very occasionally sea lions, leopard seals and elephant seals spend time on the beach. Do not approach within 5 m of any seal on land, as they can be aggressive.

Bottlenose dolphins regularly patrol the waters of the reserve and several other dolphin species have been recorded. Orca (killer whales) are infrequent visitors. Large whales that migrate along the Gisborne coast include sperm, southern right and humpback whales.

Coastal flora and fauna

Pohutukawa are the most common trees along the cliffs. Erosion is a problem for many other species and even the pohutukawa seem to hang precariously above the beach.

The beach and intertidal flats are feeding areas for red-billed and black-backed gulls. Oystercatchers, terns and little blue penguins are regularly seen, with the latter nesting in the flaxes above the beaches. Australasian gannets from the Cape Kidnappers colony to the south are regular feeders in the waters of the marine reserve.

Activities

Snorkelling
Access over the sandstone reefs is easy by entering the water north of the Pouawa River. At low tide some very large rockpools with abundant marine life can be snorkelled. This is one of the few places you can see crayfish in rockpools.

Scuba-diving
Best access is by boat around the deeper reefs not far from the shore. In spite of the limited visibility there is plenty to see. Be aware of any easterly swell and wind change.

Kayaking
The sandy beaches along the north Gisborne coast can be kayaked during offshore winds. Swells can make landing difficult at some beaches and submerged rocks and reefs hidden by the dirty water are also hazards.

Walking
Beach walk (2 hrs return)
Start from the Pouawa River end of the reserve and walk north along the sandy beach at low tide. This beach walk is around 4 km long and finishes at the Waiomoko River. On the way stop for a study of the intertidal reef platform around Pariokonohi Point, near the southern boundary to investigate the marine life. There are often seabirds on the beach and feeding close to shore.

TE ANGIANGI MARINE RESERVE

Te Angiangi Marine Reserve is an example of a fairly ordinary stretch of coastline, typical of the surrounding area. Like most of the central Hawke's Bay coast its marine life was being depleted by overfishing and collecting. Changes in this marine reserve will show how quickly such a habitat can recover.

Created: 1997

Size: 446 ha

Boundaries: Two yellow triangular markers south of the Aramoana camping ground mark the northern boundary at MHWS (the high water mark on the highest tides). When the markers are aligned, the boundary describes a north-east line that intersects a line from two more yellow triangular markers at the Ouepoto Stream. This line marks the seaward boundary out to 1 nautical mile. The southern boundary, 100 m south of Long Range Road, Blackhead, is also marked by two yellow triangular markers. The seaward boundary is 1 nautical mile from the MHWS.

How to get there: From Hastings travel south on SH 2 for about 30 km and turn off at Waipawa or Waipukurau. The drive to Aramoana or Blackhead beaches takes about 30 minutes. Access to the reserve itself is by boat, kayak or on foot. Boats and kayaks can be launched off the beach.

Best time to visit: During a westerly, when there is little swell on the east coast. Extreme low tides are the best time to explore the rock pools.

Activities: Snorkelling, scuba-diving, kayaking, walking.

Facilities: Toilets and changing rooms at Aramoana and Blackhead.

Rules: Marine reserve regulations apply (see page 12). Horses, bikes and vehicles are allowed on the sandy beach but are not permitted on the rock platform. Contact Department of Conservation's (DoC) Hawke's Bay office 06 834 3111.

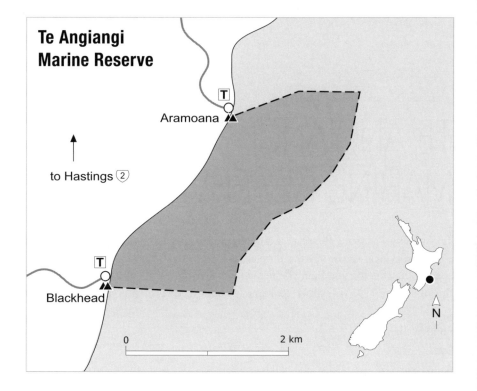

Te Angiangi Marine Reserve

History

The Aramoana–Blackhead area has been settled by Maori over hundreds of years. Ngati Kere are the local iwi (tribe) and gave the name of one of their ancestors to the reserve.

The idea of a marine reserve was originally raised in 1981 by the Pourerere Beach Improvement Association. They lobbied the Ministry of Agriculture and Fisheries (MAF) for a reserve a few kilometres north of Aramoana. Ngati Kere, Ngati Whatuiapiti and Tamatea Taiwhenua were all involved in consultations. In 1987 MAF included the proposal in a discussion paper on marine protected areas in the Central Fisheries Management Area. The Department of Conservation (DoC), formed in 1987, took over the proposal and investigated several potential sites in the area, only to meet opposition from commercial rock lobster fishers. The fishers suggested alternative locations, and the Te Angiangi site was finally decided on after support from Maori and adjacent landowners. Impacts on recreational and commercial fishers and the provision of services for visitors to the reserve were all carefully considered before the site was granted full marine reserve status.

Topography

The reserve includes an intricate reef system and large areas of sandy sea floor. Extensive mudstone platforms, interspersed with sand, line the shore. Above the reserve the hills are tertiary mudstone, which have been cleared of coastal vegetation for farming. A distinctive feature is a large pool near the centre of the reserve, known as Stingray Bay, which forms a sheltered lagoon at low tide. Below the low tide line are siltstone reefs and boulders. At the northern end of the marine reserve a boulder bank extends out to the seaward boundary where it meets the sand at around 30 m depth.

Marine life

Intertidal platforms

The intertidal reefs are covered with red, green and brown seaweeds. Clumps of brown Neptune's necklace, made up of bubble-like strands, are the most common seaweeds. They are surrounded by patches of pink coralline algae. Common grazing seashells such as cat's eyes, spotted top shells and horn shells move across the flats beside the less mobile, distinctively orange-golden limpets.

Rock pools host a variety of marine life including common triplefins, crested weedfish and juvenile spotties, the latter two among or near the seaweeds.

Under the rocks and reef underhangs near the low tide mark an increasing number of paua graze. Cook's turban shells also graze in this zone and below the low tide mark.

Subtidal reefs

The shallow areas are dominated by brown algae with several species of flapjack seaweeds the most common. Spotties, banded wrasse, marblefish and triplefins are the fish seen most often. Sea urchins graze over the reefs and turban and top shells sit among them. Crayfish are becoming more common in the cracks and under the larger rocks.

RESEARCH

DoC is currently undertaking a reef fish-tagging project. The common reef fishes of the reserve – banded wrasse, marblefish, butterfish and blue moki – have been tagged at six sites, three inside the reserve and three at nearby reefs outside the reserve. There is continued research into the movement of fishes within and between reef systems.

Common kelp gradually takes over at around 6 m and butterfish and dwarf scorpionfish are regularly seen. Light levels underwater are low and the kelp forests diminish at around 15 m. Encrusting invertebrates, like sponges and hydroids, which live under the kelp, predominate in the open. Red seaweeds cover the other available spaces on the reef.

Boulder bank

The boulder bank at the northeast end of the reserve varies between 14 and 35 m depth. The shallowest parts are very large boulders with common kelp growing across them. On the sides of the boulders colonial encrusting invertebrates dominate, with a mix of sponges, hydroids, anemones, bryozoans and compound ascidians. Nudibranchs among them are most often the gold-lined and clown species.

Schools of butterfly perch hover above the boulders. Scarlet wrasse, sea perch and red-banded perch become more common, the latter two usually sitting on the flat patches. Blue moki and tarakihi are recovering with small schools and individual fish now reasonably common. Crayfish are more noticeable with individuals of all sizes, up to 3–4 kg scattered among the boulders, often in large groups. Gurnard and goatfish feed on the sand around the boulders. Stargazers are fish that spend most of their time buried in the sand with only their eyes and mouths exposed. They wait for small fish or crabs to venture close, then erupt from the sand and swallow their prey whole. Hapuku were once referred to as common along this coast by local fishers but to date have not made a comeback.

Marine mammals

Pods of common and bottlenose dolphins visit often, while dusky dolphins and orca (killer whales) are more scarce. New Zealand fur seals, usually young males, occasionally haul out on the rocky areas along the coast. Remember to keep at least 5 m away from any seal on land.

Coastal flora and fauna

The drying intertidal platforms attract numerous feeding birds. Kingfishers hunt crabs and small fish in the pools and herons, pied stilts and variable oystercatchers also feed here. Some migrant wading birds, including eastern bar-tailed godwits, feed on the flats. The farmland behind the reserve has a few flaxes where little blue penguins nest. Australasian gannets often feed on the fish schools in the waters of the reserve.

Activities

Snorkelling
There is good snorkelling anywhere along the reef between Aramoana and Blackhead. Shelly Bay and Stingray Bay near the centre of the reserve are exceptionally good but it means a walk carrying gear.

Scuba-diving
From the beach you can explore the shallower areas, but the more interesting boulder bank is towards the outer edge of the marine reserve and easiest dived from a kayak or boat. Be aware of large swells and wind changes that can affect the coast.

Kayaking
Kayaks can be launched from either of the two beaches – Aramoana or Blackhead. Often there is a swell running which makes kayaking uncomfortable.

Walking
Coast walk (1 hr return)
With road access you can begin at either Aramoana or Blackhead Beach. Depending on the tide, you can walk along the high tide mark one way and explore the intertidal reefs when coming back or vice versa. It's about 3 km between Blackhead Beach and Aramoana Beach.

General
Longer beach walks outside the marine reserve are best done at low tide.

KAPITI MARINE RESERVE

Kapiti Island is one of New Zealand's prime wildlife sanctuaries. The marine reserve adjacent to the island is the first and only no-take marine reserve on the North Island's west coast. The marine reserve is separated into two sections, one on the exposed western side of Kapiti Island and the other between the island and Waikanae Beach.

Created: 1992

Size: 2167 ha

Boundaries: The western reserve's northern boundary extends north-west from two yellow triangular markers at Hole in the Wall Bay, out to 1 nautical mile. Its southern boundary extends seaward from a yellow triangular marker at Trig Point. The outer western boundary is defined by lining up the triangular markers at Otehape Stream and Kaiwharawhara Point and intersecting the line from the markers at Hole in the Wall Bay. The eastern reserve's northern boundary is marked by a yellow triangular marker on the northern side of the Waikanae river mouth to two triangular markers at Whakahoua on Kapiti Island. The southern boundary is from two triangular markers on the south side of the Waikanae river mouth to a green navigational buoy at Passage Rocks and then to a triangular marker at Waterfall Bay.

Getting there: Turn off SH 1 at Waikanae, 70 km north of Wellington. Public boat launching sites are at Waikanae and Paraparaumu but a four-wheel-drive vehicle is advised. Kapiti Island is 5 km off the Waikanae Coast and can be kayaked. Boat trips taking visitors to Kapiti Island leave from Paraparaumu Beach and need to be booked in advance especially during the busy summer months. A permit from the Department of Conservation (DoC) is required to land. Landing is on the bouldery beach near Rangatira Point, adjacent to the ranger's house.

Activities: Snorkelling, scuba-diving, kayaking, swimming, walking, birdwatching.

Best time to visit: Summer during calm weather.

Facilities: Kapiti Island has toilets, shelter and an information kiosk near Rangatira Point. Waikanae and Paraparaumu beaches have toilets and changing rooms.

Rules: Marine reserve rules apply (see page 12). A DoC permit is required to land on Kapiti Island. Contact DoC's Waikanae office 04 472 5821 for more information.

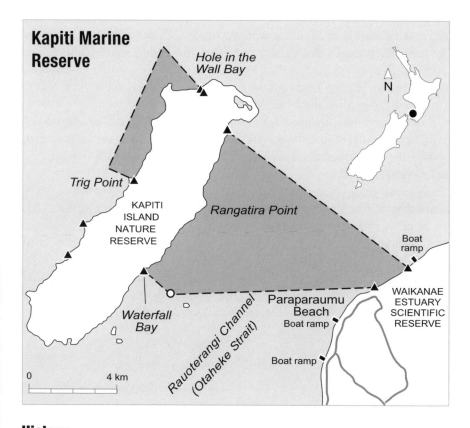

History

The island's full Maori name is Ko te Waewae Kapiti o Tara Raua ko Rangitane and is a description of the meeting place of the territories of Tara and Rangitane. According to legend, Tara was the son, and Rangitane the grandson, of Polynesian voyager Whatonga. Whatonga divided the country in two by drawing an imaginary line from Kapiti Island's southern tip to the east coast of the North Island. He gave everything above the line to his son Tara and everything below the line he bequeathed to his son Tautoki. Tautoki in turn passed the northern region on to his son Rangitane.

Rangitane, Muaupoko and later Ngati Kahungunu inhabited the island and coast. Invading Ngati Toa, led by Te Rauparaha, warred with local iwi (tribes) and took over Kapiti Island, establishing several pa in 1822. Te Rauparaha traded with European sealers and later whalers, who set up on the small islands to the southeast and along the eastern side of the island to hunt the southern right whales that moved through the channel. By 1846 whaling had become unprofitable. Te Rauparaha was removed by the Government around this time and most of the Maori left along with the whalers. Evidence of the whalers can be seen in the presence of a number of try-pots (for boiling whale blubber).

Farms were established in the 1840s and most of the island's virgin forest was burned for grazing land but farming was never more than marginal. Sheep, cattle, goats, pigs, deer and possums were introduced.

Kapiti Island was designated a nature reserve in 1897 and eradication of pests began in 1918. In 1996 the last kiore (Polynesian rats) were eliminated, making the island predator-free. A small area of 13 ha around Waiorua Bay is still privately owned.

Concern over deterioration of the marine environment led to two boating clubs establishing voluntary reserves on the eastern side of Kapiti. During the 1980s the Ministry of Fisheries commissioned an investigation into the establishment of a marine protected area, and in 1987 the newly formed Department of Conservation began to consider a marine reserve around Kapiti Island. A 1989 discussion paper received overwhelming public support for a marine reserve that would complement the adjacent Waikanae Estuary Scientific Reserve and the Kapiti Island Nature Reserve. The area was accorded full marine reserve status in 1992.

Topography

Kapiti Island is believed to have been part of a land bridge that connected the North and South Islands. It is approximately 8 km long and 1.5 km wide, and made up mainly of greywacke and argillite rock. At the northern end there are some sandy beaches, but most are pebbles and small boulders.

The western side of the island is dominated by high cliffs and ridges running the length of the island. The eastern side is shaped by valleys formed by the island's nine major streams and slopes more gently. The highest point on the island is Tuteremoana at 524 m.

Two currents, the warm Westland Current that travels north around Farewell Spit and the cooler D'Urville Current, converge south of Kapiti and influence the marine life in the reserve. The area off Waikanae and Paraparaumu is the remnants of a drowned river delta.

Marine life

The waters around Kapiti Island include four different habitat zones, typical of the region, but only three are included in the marine reserve.

Channel

The Rauoterangi Channel between the Waikanae coast and Kapiti Island reaches 80 m in depth and is mainly silt, sand and gravel washed down from rivers and streams. Seaweeds are uncommon but there are large worm beds, providing food for gurnard, goatfish and skate. Eagle rays leave large depressions in the sand as they dig for crabs and seashells. The area experiences strong tidal currents.

Rocky reefs and boulders

The eastern side of Kapiti Island from Whakahoua to Waterfall Bay changes to a boulder bottom with rocky reefs intersected by sand. Although sheltered from prevailing westerly winds, the area experiences strong tidal currents. Sea urchins, sea stars, solitary corals, anemones, octopus and crayfish frequent the boulders. Seaweeds cover much of the shallow reefs with the flapjack variety being the most common. Grazing molluscs include paua, tiger and Cook's turban shells. Small red crayfish form colonies in the cracks and under the larger boulders.

Spotties are the dominant fish around the weedy areas with some butterfish, leatherjackets, kelpfish, marblefish and banded wrasse. Schools of silver drummer feed on the seaweed forests and in summer large schools of jack mackerel move in and feed around the shallower reefs. Kahawai schools feed near the surface and attract terns. Eagle rays feed on the sand, along with goatfish and blue cod. Below 15 m schools of tarakihi are common. Concentrations of unusual, free-living, calcareous algal growths known as rhodoliths occur in the area.

RESEARCH

An initial National Institute of Water and Atmospheric Studies (NIWA) 1992 study of marine organisms was carried out prior to the reserve's creation. Between 1998 and 2000 a study by Victoria University compared size differences and abundance of 34 species, including fish, invertebrates and seaweeds, at sites inside and outside the marine reserve. Their findings were then compared with those from the 1992 NIWA study. No major changes were found between the data from this survey and the 1992 NIWA survey.

A 1999–2000 survey by NIWA did find a significant increase in crayfish numbers within the marine reserve compared with fished areas surveyed. No significant size difference was noted. Also, butterfish numbers and sizes on the western side of Kapiti Island had increased compared with areas outside the reserve.

Blue cod monitoring by Victoria University in 2003–2004 used a wide range of survey methods. The data showed that blue cod were larger and more abundant in the reserve than outside.

Western reefs

The western side of the island from Hole in the Wall Bay to Trig Point has extensive boulder reefs, extending down from rocky headlands and cliffs. The water is clearer than on the eastern side of the island but the reserve is more exposed to prevailing wind and wave action.

Kelp forests dominate the rocky reefs and snapper and red moki are often seen from the shallows to the deeper areas. Large octopus glide over the rocks and feed on crayfish and crabs.

Below the seaweed zone, sponge gardens become dominant and schools of butterfly perch, tarakihi and occasionally snapper feed. Sea and banded perch sit on the ledges in the deeper reefs. Some migrants from warmer waters, such as spotted black grouper, have been seen. Cold-water fish like magpie perch and large blue moki from the open ocean also visit. In summer predators such as kingfish feed among the schooling fish, and blue and bronze whaler sharks pass through.

Marine mammals

New Zealand fur seals have established a winter hauling out ground at Arapawaiti Point. They often interact with divers by moving close and swirling around them near Hole in the Wall Bay. They are mainly young males that feed on squid and lanternfish in the deep water to the west. Bottlenose dolphins and orca (killer whales) visit Kapiti. Southern right whales migrate from the subantarctic islands and are occasionally seen near Kapiti Island. Humpback whales pass as they migrate from the Antarctic to mate and breed in the Pacific Islands.

Coastal flora and fauna

Kapiti Island has been predator-free since 1996 and has increasing numbers of birds, reptiles and insects. The bush has recovered from years of grazing and covers almost all the island. In the early 1900s little spotted kiwi, which were almost extinct on the mainland, were introduced to the island, and have done very well there. In fact, the island now has New Zealand's largest little spotted kiwi population.

Colonies of red-billed and black-backed gulls live along the rocky coasts. Little blue penguins nest on Kapiti Island and their burrows are sited among the flax and bushes near the beach. Black and little shags nest in the trees and fish offshore. Rare reef herons hunt in the rock pools and variable oystercatchers wander the beaches and intertidal reefs.

Offshore are flocks of sooty shearwaters, which nest on the western cliffs. White-fronted terns, Australasian gannets, Caspian terns, diving petrels and fluttering shearwaters feed around the island.

Activities

Snorkelling
Best snorkelling is close to Kapiti Island especially around the Hole in the Wall Bay and Arapawaiti Point, accessed by boat. A highlight is a New Zealand fur seal encounter with one of the animals that live here.

Scuba-diving
The reefs on the western side of the island are the most interesting for divers. The only access is by boat and often the most sheltered spot is north of Arapawaiti Point. Currents and limited visibility make the western side less attractive.

Kayaking
Launch kayaks off the beaches at Paraparaumu or Waikanae. The 5 km channel crossing can be rough, especially if the wind opposes the tide. A permit is required to land on the island. It's a long paddle around the island and for experienced kayakers only. Close to the western shore of Kapiti you may see and hear some of the endangered bird life. Little blue penguins sit on the surface during the day and allow kayaks to approach closely before they dive away.

Walking
The public tracks are well signposted near the landing site at Rangatira Point.

Trig Track (3 hrs return)
This track climbs through regenerating native bush and is the recommended track up to the Tuteremoana Lookout at 524 m. It's very steep in places and can be muddy and slippery in wet weather. The lookout tower at the top offers magnificent coastal views over the marine reserve. It joins the Wilkinson Track three-quarters of the way up. If you stop and wait at any point along the way, you may catch a glimpse of the endangered birdlife.

Wilkinson Track (3 hrs return)
This track also leads to the Tuteremoana Lookout. This is the easier of the two tracks and the recommended downhill route from the summit as it is not as steep, though again the track can be muddy and slippery in wet weather.

North Track (2–3 hrs return)
This easy track follows the coast to Waiorua Bay. The track detours onto the beach, past 13 ha of privately owned land, and on to Okupe Lagoon. You'll see lots of bush birds on the way and large numbers of water fowl on the lagoon.

WESTHAVEN (TE TAI TAPU) MARINE RESERVE

This marine reserve is located at the southern end of a beautiful, largely unmodified South Island estuary, surrounded by regenerating natural forest. It is an important habitat for wading birds, bordering the Westhaven Wildlife Management Reserve.

Created: 1994

Size: 536 ha

Boundaries: The marine reserve includes all the tidal sand flats and channels of Whanganui Inlet south of a straight line between two triangular markers at both Melbourne Point and the closest headland of Westhaven Scenic Reserve, and all tidal areas upstream of causeways along Dry Road, southwest of, and including, the Wairoa River. Excluded is a small area around the Mangarakau Wharf where fishing is still permitted.

Getting there: From Collingwood, follow the main road to Farewell Spit and turn left just north of Pakawau. The Whanganui Inlet is just past the turn-off to Kaihoka Lakes. Dry Road winds around the inlet to Mangarakau. The marine reserve begins at the Wairoa River mouth and is accessible by boat, kayak and on foot. Small boats and kayaks can be launched at the Mangarakau Wharf or 200 m past the Kaihoka Road turn-off.

Activities: Kayaking, walking, swimming.

Best time to visit: High tide is the best for swimming and kayaking.

Facilities: None.

Rules: Marine reserve regulations apply (see page 12). Contact the Department of Conservation's (DoC) Takaka office 03 525 8026. Domestic animals are not permitted in the marine reserve.

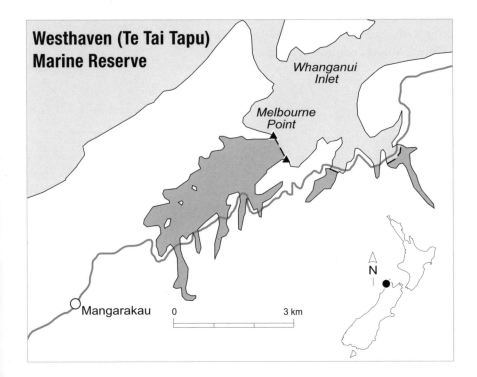

Westhaven (Te Tai Tapu) Marine Reserve

History

The Whanganui Inlet was an important source of seafood and there is evidence in the area of Maori occupation. By the time European settlers arrived, the number of Maori inhabitants was low, probably due to the battles with other Maori who moved through the area during the 1820–1830s. Several pa sites are dotted around the estuary along with mounds of midden material.

The coastal forest bordering the inlet was clearfelled and selectively logged for rimu, kahikatea and beech. Large areas of flax were cut and processed, and coal and gold were mined for a brief period around 1900. Most of the area has since regenerated.

In 1987 the Ministry of Agriculture and Fisheries suggested protection for the area. This proposal was followed up by DoC who completed a survey of the inlet's marine life in 1989–90. The results showed the inlet was indeed worthy of marine protection and an application for a marine reserve was lodged. Local consultation supported some marine protection while keeping traditional and recreational fishing in part of the inlet. To accommodate this, the Westhaven Marine Reserve was created in 1994 protecting all the marine life within its boundaries. A month later the Westhaven Wildlife Management Reserve was created covering the rest of the inlet and protecting the wildlife and habitats, but allowing fishing and gamebird hunting.

Topography

The Whanganui Inlet is an enclosed, drowned river valley and the only large estuary of its kind on the South Island's west coast. Almost all the catchment of the marine reserve is covered in dense forest, which gives way to salt marsh at the water's edge. Much of the inlet is exposed at low tide. Water from the narrow sea entrance feeds deep channels running northeast and southwest.

Marine life

The estuary is dominated by large tidal flows, which leave most of the sand flats uncovered at low tide. Eel grass beds cover much of the area from mid tide, creating an important nursery area for snapper, flounder, kahawai and whitebait. Approximately 30 species of marine fish and 163 species of invertebrates have been recorded in the inlet. The invertebrate figure is the highest of any South Island estuary. Large numbers of freshwater fish migrate through the estuary as part of their life cycle. Banded kokopu, inanga, red-finned bullies and long-finned eels are the most common of these migratory native fishes.

Mudflats

Soft mud covers the higher intertidal areas up to the salt marshes that surround the reserve. Tidal channels, with their source in the small streams that enter the estuary, meander across the mudflats. The mudflats themselves are pockmarked with holes – the burrows of several species of mud crabs. The crabs do not venture far from their holes and will scurry back at the first sign of danger.

The most common seashells are mud snails, which feed on bacteria in the mud and leave distinctive trails of raised mud. Mud whelks and the much narrower

RESEARCH

A baseline survey of estuarine habitats and communities (including detailed habitat mapping) was undertaken in 1988 and 1989, prior to the creation of the marine reserve. There have been no follow-up surveys undertaken, apart from colour aerial photography of the entire inlet in 2001.

horn shells are also dotted around, the former often leaving a long groove in the mud as they travel. Small patches of eel grass exist towards the mid tide areas.

Estuarine triplefins and juvenile flounder can be seen in the tidal channels. During spring as the tide rises schools of whitebait, the juvenile forms of native galaxid trout, move up into the fresh water above the tidal influence.

Eel grass beds

The mud gives way to fine sand, which becomes more solid at the half tide mark, and eel grass beds become dominant. Patches of cockles mix with smaller bivalves like nut shells. Cockles are abundant in the area northwest of Melbourne Point. Odd, footprint-like marks on the sand are from the feeding siphons of wedge shells; when submerged, the siphons are raised above the sand to extract plankton and exhale water. Mud and green harbour crabs are quite common and their burrows are scattered among the eel grass. Snapping shrimps, with their outsized left claws, also live here. Mantis shrimps live in circular-holed burrows and can be enticed out by leaving scraps of food on the up-current side.

As the tide moves across the flats, larger flounder and yellow-eyed mullet move into the shallows to feed over the cockle beds. Pipi beds are near the low tide area. Spotted and mud whelks, along with brightly coloured inflated cushion sea stars, are the major predators.

Rock platforms

The area around Melbourne Point is the best example of rock platform habitats in the reserve. Cat's eyes are the most common gastropods and blue and ribbed mussels the most common bivalves. Other seashells include dark rock shells, periwinkles and oysters.

Sea lettuce and Neptune's necklace seaweeds are scattered across the platforms. In the cracks above the high tide mark are large purple shore crabs. Red rock crabs live close to the low tide mark in rock pools, along with cushion stars and common sea urchins. Several species of triplefin also inhabit the rock pools.

Subtidal channels

Large schools of kahawai are quite common in summer and autumn, feeding in the channels. Pipi beds extend into this zone and larger trough shells are common, along with morning star shells. Deep holes in the sand, up to half a metre across, are from eagle rays, made as they forage for shellfish and crabs.

Large paddle crabs feed in the channels, which are thought to be major nursery areas for snapper and flounder. During prolonged periods of heavy rain, run-off affects the harbour and it is thought some fish species move out of the estuary until the salinity level returns to normal.

Coastal flora and fauna

The tidal flats in the estuary are one of the most important regions in the Nelson area for wading birds. A significant proportion of the region's banded dotterels breed here. Large numbers of eastern bar-tailed godwits, knots and oystercatchers feed on the flats and in the eel grass beds. The salt marsh areas provide cover for the banded rail and this inlet is the only place on the west coast of the South Island where they are found. Bitterns, fernbirds and spotless crakes are also found here. Kingfishers and white-faced herons feed on the sand flats.

Little shags and black-backed seagulls are the most numerous seabirds nesting around the inlet. Australasian gannets, white-fronted and Caspian terns, and red-billed gulls feed in the inlet but probably breed on Farewell Spit.

Activities

Kayaking

This is the best way to explore the marine reserve. Kayaks can be launched at various places along Dry Road, which runs alongside Whanganui Inlet. As the tide rises various estuaries and channels can be explored. The largest is the Wairoa River which can be followed for several kilometres inland alongside native forest. Best to enter about an hour before high tide.

Walking
Knuckle Hill Walk (45 mins return)

Turn off Dry Road, just past Banjo Creek, onto Knuckle Hill Road and it's a 5 km drive over a fairly rough road to the car park. The track is steep but worth the walk to the summit at 506 m, where there are views all round including over the Whanganui Inlet.

TONGA ISLAND MARINE RESERVE

The reserve borders the Abel Tasman National Park with its excellent coast track and is one of New Zealand's most popular kayaking destinations. A highlight of the marine life is the colony of New Zealand fur seals on Tonga Island.

Created: 1993

Size: 1835 ha

Boundaries: The offshore boundary runs due east for 1 nautical mile from the headland that separates Mosquito Bay and Bark Bay. The boundary then runs 1 nautical mile from the coast and seaward from Tonga Island. The northern point is 2340 m northeast of Awaroa Head, in line with Cave Point. Yellow triangular markers on shore mark the northern and southern boundaries. A directional orange flashing light on the southern marker is visible as soon as you enter the reserve from the south.

Getting there: There is no road access. Foot access is along the Abel Tasman Coast Track from the north or south. The nearest car park is at Awaroa Bay, 33 km from Takaka. Boat ramps are at Totarunui, Tarakohe, Kaiteriteri and Marahau. Best place for launching kayaks is Marahau. Regular water taxis service the coast from Kaiteriteri and Marahau and stop at all beaches as required.

Activities: Snorkelling, scuba-diving, kayaking, walking, swimming.

Best time to visit: The area can be visited all year round, depending on weather conditions. During summer, strong sea breezes occur almost daily, affecting visitors by boat or kayak. There is little shelter along some rocky sections where the current opposing the tide can make conditions unpleasant. During winter there are often long periods of calm.

Facilities: There are campsites and toilets along the Abel Tasman Coast Track. You must have hut or camp passes, available from the Department of Conservation (DoC). Contact DoC's Motueka office on 03 528 9117 or Takaka on 03 525 8026 for further information.

Rules: Marine reserves regulations apply (see page 12). Marine mammal regulations prohibit anyone approaching within 5 m of a New Zealand fur seal on land. Landing on Tonga Island is not permitted.

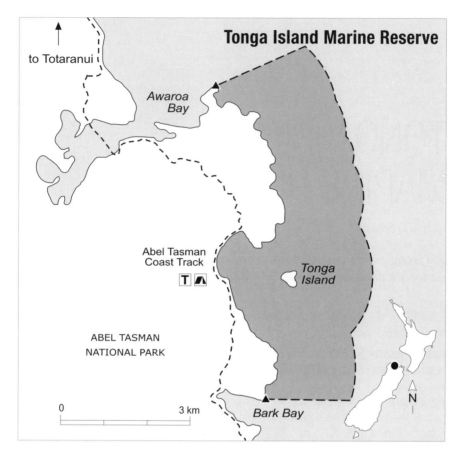

History

Maori have occupied the coast for at least 500 years, gathering the abundant fish and shellfish found here. Ngati Tumatakokiri inhabited the coast when the first European explorer Abel Tasman sailed into Golden Bay in 1642. In an attempt to land, four of his crew were killed by Maori and Tasman named the spot Murderers' Bay. Ngati Tumatakokiri were attacked by invading tribes in the late 1700s. Major warfare between tribes in the early 1800s resulted in virtually no Maori occupation by the time early European settlers arrived in the 1840s. The bay was renamed after gold was found in the area. Abel Tasman National Park was created in 1942 to mark the 300th anniversary of Tasman's visit.

The need for some form of protection became apparent as fishing pressure caused a decline in fish numbers in the Tasman Bay area. In 1990 DoC studies confirmed the scarcity of fish life. Several options for a marine reserve were considered, with the Tonga Island area widely supported as the best option.

Topography

The marine reserve is typical of New Zealand's northernmost sheltered granite coastline. The beaches are coarse, golden sand surrounded by bouldery headlands and reefs. The boulders change to a sand-mud sea floor at depths of 6–10 m. Tasman Bay is sheltered from ocean swells, but is influenced by wind-generated waves, often from sea breezes. A change in wind direction can quickly increase or decrease wave height.

Marine life

The Nelson coast is protected from southerly storms, which pound the northern side of Cook Strait. The main seaweeds are brown flapjacks and encrusting red coralline algae. Blue mussels, barnacles, cat's eyes and other grazing molluscs share the habitat with common sea anemones and sea stars.

Boulder reefs

Like much of the coast, large boulders dominate the shallows. Above the low tide mark, limpets, nerita and cat's eyes are the most often seen seashells. Around the low tidemark, the boulders play host to small stubbly seaweeds, along with tube worms, cushion stars and limpets. Below low tide the lack of the usual seaweeds is evident and large groups of sea urchins cover the boulders. Triplefins are abundant and dozens of these 5–10 cm fish dart around on the boulders.

A layer of silt covers much of the coast and is stirred up by wave action from west and northwesterly winds often creating poor underwater visibility. During calm conditions visibility can exceed 15 m.

The boulders give way to sand between 5 m and 12 m. The lack of large kelp is obvious but there is plenty of other marine life dominated by sea urchins, grazing limpets, Cook's turban and green top shells. Nudibranchs include 15 cm Wellington nudibranchs, with their yellow or brown warty bodies. Grey flatworms, similar in shape to nudibranchs, are also common, often crawling across the boulders.

RESEARCH

Regular studies monitor the changes in marine life in the reserve. Data shows that crayfish numbers have increased notably since 1993. Snapper have been tagged and ongoing studies will show any increase as well as their movements.

Patches of common sea anemones live on the boulders among sponges, the latter more common in the shaded areas.

Two species of cushion sea stars live alongside larger spiny sea stars. The feeding methods of the bright orange-coloured, inflated cushion stars are worth observing. The cushion star raises its body, leaving tempting gaps underneath. When a triplefin or other small creature darts into the gap, the cushion star drops, grasping its prey with its sucker feet.

Boulder edge and sand flats

Several species of triplefins, mostly common or variable varieties, flit across the boulders. Above them, the most common fish are banded wrasse, which move in close to investigate diving intruders. Often as many as six wrasse follow divers. Juvenile spotties shelter in the cover of flapjack seaweeds, while the adults follow in the wake of divers, ready to pounce on any prey that has been disturbed. Blue cod, leatherjackets, tarakihi and occasional red moki move around above the boulders. Near the sand, the boulders become larger with patches of encrusting sponge life. The gaps between the boulders are good habitats for red crayfish although most are small to medium size.

Sea cucumbers lift particles of sand to their mouths with their oral tentacles. The sea cucumber's digestive system removes any bacteria from the sand particles, which emerge from the rear of the sea cucumbers as clean sand. Black, slug-like shield shells, up to 15 cm in length, crawl across the sand. The shell itself, shaped like a Roman shield, is usually hidden under the folds of skin on the upper parts and usually only emerges if the animal is touched.

Blue cod watch for anything edible that may be dislodged from the sand, and small schools of juvenile tarakihi live near the deeper parts of the reserve, especially near the northern boundary. Snapper were once common but since being overfished have not made a return, apart from isolated individuals. The disappearance of snapper from the waters around the coast was one of the reasons behind local support for the establishing of a marine reserve.

In the sandy areas, scallops and large horse mussels feed on plankton carried by the currents. Predatory gastropod molluscs, Arabic volutes and ostrich foot shells feed over the sand, but remain buried during the day. Large paddle crabs leap out from the sand in a threat display before racing away to rebury themselves. Schools of goatfish use the barbels set below their mouths to fossick in the sand for worms and crustaceans.

Creek entrances and estuaries

The small creeks and estuaries have some shellfish beds with cockles the most common bivalve. Mud snails and whelks leave their distinctive trails in the sand.

Small crabs scuttle about and dive under the sand or into a hole if disturbed. Schools of yellow-eyed mullet move up the estuaries at high tide and in summer the native galaxids, better known as whitebait, move up into the wetlands behind the estuaries to breed.

Marine mammals

New Zealand fur seals breed on Tonga Island and are regular visitors to the surrounding coast. They should not be approached on land, as they can be aggressive, especially if their path to the sea is blocked. In the water they are agile and can be seen by kayakers, divers and snorkellers..

Dolphins, mostly dusky or bottlenose, are frequent visitors and the smaller and much rarer Hector's dolphins are occasionally seen in summer.

Coastal flora and fauna

The coast is dominated by beech forest and bellbirds make their presence known with morning and evening calls at the campsites. The bush, which reaches to the shore around most of the coast, is a nesting area for little blue penguins that feed at sea during the day. Australasian gannets, from the colony at Farewell Spit, feed in the area, along with shags and fluttering shearwaters. There is a small estuary behind Onetahuti Beach where several species of wading birds, including white-faced herons and oystercatchers, feed as the tide drops.

Activities

Snorkelling

During calm periods, the water in the marine reserve can be very clear. While snorkelling around the reefs is interesting the highlight is an encounter with a New Zealand fur seal. The seals come in quickly and check out snorkellers. They roll and turn and if you copy their actions they will try to out-do you, leading to a memorable encounter. Access is easy from most of the coast, but a boat or kayak is handy to explore the outer islands and reefs. Stay close to the coast or rocks, as there is quite a bit of boat traffic, especially in summer.

Scuba-diving

The best scuba-diving is around Tonga Island and among the reefs along the northern coast of the marine reserve. Although a beach entry is possible at Onetahuti, it is best to dive from a boat or kayak. If you are lucky you may see a fast-moving fur seal anywhere along this coast.

Kayaking

The Abel Tasman coast is one of New Zealand's most popular kayaking destinations. Guided kayaking trips start from Marahau Beach and explore the coastline over two or three days. Experienced kayakers can hire kayaks or take their own. The sea kayaks have room to store camping gear and food. Dolphin encounters are common and they'll often leap out alongside your kayak. Fur seal encounters are more likely around Tonga Island. Campsites, bordering the reserve, are at Mosquito Bay, Tonga Quarry and Onetahuti Beach. Many kayaking trips finish at Onetahuti Beach and the return trip, including kayaks, is made by water taxi.

Walking
Abel Tasman Coastal Track

The Abel Tasman Coastal Track covers 51 km between Marahau and Totaranui and passes the marine reserve. To get to the marine reserve it's around a day's walk from Awaroa Bay to Onetahuti Beach and two days from Marahau. Campsites bordering the reserve are at Onetahuti and Tonga Quarry. Packs and tents can be ferried by water taxi to campsites. You have the option of walking all or part of the track or combining your walk with a kayaking section.

Above: *Marine reserve rules and boundaries are spelt out at Te Tapuwae o Rongokako.*

Below: *Local red crayfish numbers have recovered within Te Tapuwae o Rongokako Marine Reserve.*

Top: *A low-tide beach stroll at Te Angiangi Marine Reserve offers superb Hawke Bay vistas.*

Above: *Red gurnard live on sandy areas where they feed on invertebrates.*

Top: *New Zealand's most commonly seen sea slug, the clown nudibranch, can be found in rock pools.*

Above: *Seaweeds offer a safe haven from predators for camouflaged fish such as this crested weedfish.*

Above left: *Jewel anemones cover subtidal rocky areas at Sugar Loaf Islands (Nga Motu) Marine Protected Area.*

Below left: *Sand anemones come in a variety of colours and occur near Kapiti Island.*

Above right: *Wandering anemones are found right around the New Zealand coastline.*

Above left: *Whanganui Inlet is an important habitat for freshwater fish and wading birds.*

Below left: *Large Wellington nudibranchs graze the boulders in Tonga Island Marine Reserve.*

Above: *Tonga Island, off the Abel Tasman Coast, is home to a colony of New Zealand fur seals.*

Right: *Jewel anemones nestle among seaweeds in Long Island–Kokomohua Marine Reserve.*

Above: *Blue cod numbers have increased inside Long Island–Kokomohua Marine Reserve.*

Below: *Yellow-foot paua graze on algae-covered rocks at Long Island–Kokomohua.*

HOROIRANGI MARINE RESERVE

The Nelson Boulder Bank, although a significant geological feature, gives no clue to the wealth of colourful marine life living less than 15 m underwater. Sponge gardens cover the deeper reefs of the reserve in a variety of shapes and colours not seen elsewhere in Tasman Bay.

Created: Approved in 2005 but not gazetted

Size: 904 ha

Boundaries: The southern boundary extends from a yellow triangular shore marker at Glenduan to a buoy 500 m offshore, to a second buoy, 1 nautical mile offshore, which marks the outer boundary. The northern boundary extends from a yellow triangular shore marker at Ataata Point, north-north-east 300 m to a buoy then north-west to the outer buoy, 1 nautical mile offshore. The seaward boundary follows the coastline 1 nautical mile from shore.

Getting there: From Nelson, take SH 6. The turn-off to Glenduan (the southern end of the reserve) is 11 km north of Nelson. Access to the northern end of the reserve is from Cable Bay, 22 km from Nelson. The nearest boat ramp is at Nelson. Kayaks are best launched at Cable Bay.

Activities: Snorkelling, scuba-diving, swimming, surfing, kayaking, walking.

Best time to visit: Summer during calm days and when the wind is from the east.

Facilities: There are public toilets at Cable Bay and Glenduan.

Rules: Marine reserve regulations apply (see page 12). Contact Department of Conservation's (DoC) Nelson office on 03 546 9335.

History

Maori occupation in the region is believed to date back more than 1000 years. From the archaeological evidence, it would appear that early Polynesian settlers visited the Nelson Boulder Bank to collect hammer stones. Despite the fact that

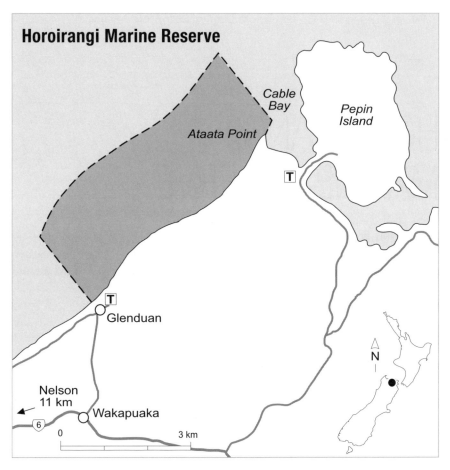

Maori lived at Glenduan and Cable Bay on both a permanent and seasonal basis, there are no particular sites of archaeological interest bordering the reserve. European settlement in the 1840s was also at Glenduan and Cable Bay.

As fishing, spear-fishing and collecting escalated during the 20th century it placed the coast's marine life under pressure. Commercial fishing records show that large schools of snapper spawned in Tasman Bay during the summers, but were severely depleted by trawling activities in the 1970s.

The idea of a marine reserve north-east of Nelson was proposed in the early 1980s by the Ministry of Agriculture and Fisheries (MAF). Two areas, from Delaware Bay to Pepin Island, and from Cable Bay to Glenduan, were identified as suitable. Most returns from a 1990 questionnaire were in favour of a marine reserve at north Nelson, but no further action was taken until 1994, when Nelson Forest and Bird contracted the National Institute of Water and Atmospheric Research (NIWA) to do a scientific investigation of the coast.

Ngati Tama, the local iwi (tribe), suggested that the area south of Cable Bay was most suitable for a marine reserve and requested a taiapure (customary fishing area) to the north of the proposed site. Three sites were listed in a public discussion document released in 1996 and three-quarters of submissions favoured a marine reserve somewhere on the north Nelson coast. The Glenduan–Ataata Point option was chosen as the most appropriate site for a marine reserve in the eastern Tasman Bay. The park was named Horoirangi, which is the name of a Maori ancestor from the area, and the name of the highest peak overlooking the marine reserve.

Topography

The marine reserve is along the northern end of the 13.5 km Nelson Boulder Bank. The boulders are the result of erosion from the coastal hills between Glenduan and Ataata Point. The boulders are a mix of granite, granodiorite and andesite, commonly known as 'Tasman Intrusives'. The stones and rocks are small at the high tide level and gradually increase in size to large boulders at the low tide level. Silt from run-off and rivers covers the Tasman Bay seabed and there is little wave action to dissipate it.

Marine life

As the boulders along the intertidal area are constantly on the move, they do not tend to attract much marine life. Silt covers much of the area after heavy rain and underwater visibility is usually not good.

Boulder habitats
Some sea snails exist here, along with chitons and limpets. Window oysters (jingle shells) with their opaque, apricot-coloured shells are attached to some surfaces, plus a few small mussels. The largest seaweeds present are two species of flapjack seaweed, which extend from low tide to around 6 m depth.

RESEARCH
Numerous scientific and educational bodies have planned baseline studies both in the marine reserve and the proposed taiapure adjoining it. They include the Nelson-based Cawthron Institute, DoC, NIWA, MAF, Nelson Polytechnic and many local schools.

Cook's turban, green trochus and cat's eye sea snails are relatively common, grazing on the stubbly weeds that grow on the boulders.

Urchin barrens become apparent between 6 m and 8 m depth on the boulders. Coralline algae covers the rocks down to 10 m and sea urchins are very common along with turban shells, cushion sea stars and sea squirts. Cushion stars ambush their prey by extending their bodies upwards, then dropping quickly on small fish or crabs which venture underneath. Smaller cushion stars and spiny sea stars are also present. Large, knobbly, Wellington nudibranchs are the most often seen of several species of sea slug as they graze on the boulders. Some paua are present and large wandering anemones cling to the boulders.

Sponge gardens

The sponge gardens begin at 10 m depth and continue down to 15 m on the boulder reefs, which extend to 150–400 m out from shore. Species include large finger sponges, golfball and red, yellow, orange and grey encrusting sponges. These sponge gardens in turn provide important habitats and shelter for juvenile fish. Several species of triplefin dart among the sponges and tarakihi, blue cod, leatherjackets, red moki and banded wrasse are seen, although spotties are the most common fish. Scattered among the sponge gardens are colonies of lace coral bryozoans, typical of the large beds that once existed in the Nelson area.

The boulders sit on a sand/gravel mix, which changes to silty mud between 12 m and 15 m. Heart urchins, sand dollars, worms, gastropod molluscs, mantis shrimps, dog cockles, paddle crabs and small scallops are resident in the mud. Sea cucumbers work their way through the sediment, extracting nutrients from it and processing the clean sediment. Gurnard, goatfish, sole, opalfish, red cod, dogfish and eagle rays hover over the mud. Below 20 m there are heart urchins, hermit crabs, horse mussels, morning star shells, silky dosinia shells and brittle stars.

Northern reefs

Ataata Point, at the northern end of the reserve, has rocky reefs and outcrops mixed with boulders. Large patches of common anemones cover the rocks, and sponges and colonial corals are common on the vertical rock walls and rocky underhangs. Brachiopods – red two-part shells known as ladies' toenails – form dense coverings in the same area, usually with coatings of coralline algae. Window (jingle oysters) are very common. Small clumps of green-lipped mussels and paua cling to the rocks. Reef fish, with the exception of spotties, are not found here in any great number. Snapper, butterfish and marblefish are currently rare. Jack mackerel are the most abundant fish, with schools of up to 200 found close to the rocks, especially in late summer.

Marine mammals
Bottlenose dolphins visit occasionally, and orca (killer whales) less often. Juvenile New Zealand fur seals have been seen on the rocks around Cable Bay in winter.

Coastal flora and fauna
Colonies of up to 1000 spotted shags nest on the rocky cliffs at Ataata Point and Pepin Island. Most of the hillsides behind the reserve have been modified by farming, although significant areas of indigenous forest remain in the gullies north of Glenduan.

Activities

Snorkelling
Though visibility is limited, you can snorkel almost anywhere along the reserve. Ataata Point is the most interesting area; enter the water at Cable Bay.

Scuba-diving
It's not easy to enter the water across the slippery boulders with scuba gear and diving is recommended from a boat or kayak. The sponge gardens at the deeper edge of the boulder bank make an interesting dive.

Kayaking
Kayaks can be launched or hauled out at Glenduan or Cable Bay. Care is needed during low tide when the lower, slippery boulders are exposed and when a westerly wind creates onshore waves. From Nelson it is an 11 km paddle alongside the boulder bank to Glenduan.

Walking
Boulder Beach Walk (2–3 hrs return)
You can begin this walk from either Ataata Point or Glenduan, when the tide is falling. Good shoes are needed on the boulders. After heavy rain, some of the streams washing into the reserve may be impassable. An alternative is to walk one way along the boulders and return via the Cable Bay Walkway.

Cable Bay Walk (2.5–3 hrs one way)
You can begin at either Cable Bay or Glenduan. The track has some steep sections and passes across private farmland and through covenanted coastal forest. The track is closed during lambing and calving.

LONG ISLAND– KOKOMOHUA MARINE RESERVE

Located at the entrance to Queen Charlotte Sound, Long Island–Kokomohua Marine Reserve is a great example of the impact full protection can have on the marine life in an area. Many varieties of fish and crayfish have returned after their numbers were severely depleted by commercial and recreational fishing. A highlight for divers is an encounter with the large blue cod, which appear out of nowhere, grabbing at trailing dive straps or even fingers.

Created: 1993
Size: 619 ha
Boundaries: The offshore boundary extends 0.25 nautical miles around Long Island, the Kokomohua Islands and the unnamed but charted rock northeast of the Kokomohua Islands.
Getting there: The marine reserve is 35 km from Picton in Queen Charlotte Sound and is accessible by sea only. The reserve is best viewed by boat or kayak as foot access around the shore can be difficult. Boaties need to be wary of unmarked reefs and strong currents, especially around the northern end, and should be aware that northerly winds can quickly whip up seas. Long Island has no jetty, and the easiest landing is on the boulder beach on the south-western shore. Landing is also possible on the eastern side of Long Island. The nearest boat ramps are at Picton and Waikawa.
Activities: Snorkelling, scuba-diving, kayaking, birdwatching.
Best time to visit: Anytime during calm weather.
Facilities: None.
Rules: Marine reserve regulations apply (see page 12). Contact the Department of Conservation's (DoC) Picton office 03 520 3002 for more information. No domestic animals are permitted on Long Island.

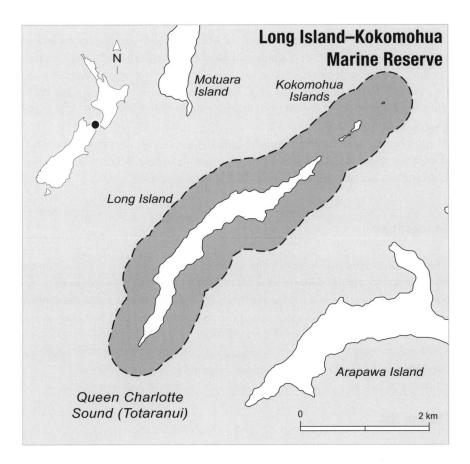

History

Long Island was known to local Maori as Hamote. The island was unoccupied when explorer James Cook landed there in 1770, though he did note an abandoned fortified pa that had evidently resisted heavy attack. Cook returned in 1773 and spent several days planting a garden on the southern tip.

In the early 1850s Captain Arthur Wakefield noted that the island was heavily fortified, but since then the fortifications and pa site have vanished. It is not known when Maori left the island or the reason why.

Early attempts to farm the island in the late 1800s failed and by 1925 it had reverted to fern and scrub. The island became a scenic reserve in 1926 and the last of the stock was removed in the 1930s.

During World War II a signal station, wharf tramway and barracks were built at the northern end of Long Island and an anti-submarine defence station on the southeastern side. A few ruins remain amongst the regenerated native bush.

During the 1980s, fishing pressure in the Marlborough Sounds caused a decline in stocks of fish and other commercial catches, especially blue cod, tarakihi, grouper, crayfish and paua. This decline was noted by members of Marlborough dive clubs who voluntarily stopped taking fish from around Long Island in 1989. They began promoting the idea of a marine reserve and discouraged fishing and collecting from the area.

After public consultation, the clubs, aided by the Department of Conservation (DoC), applied for marine reserve protection and in 1993 Long Island–Kokomohua became the South Island's first marine reserve.

Topography

The Marlborough Sounds were formed from a network of drowned river valleys after the last ice age. The marine reserve is typical of the outer sounds in its range of habitats. They include rocky northern reef areas around the Kokomohua Islands and the more sheltered area on the southwestern side of Long Island. The boulder bank on the southwestern shore of Long Island is a formation known as a cuspate foreland – a distinct triangular flat area formed as opposing currents meet and pile boulders along a shoreline.

Marine life

Much of the island's coast is dominated by rocks and boulders, except for the northern end where seaweed-covered reefs give way to sandy mud at 15 m.

RESEARCH

Monitoring of fish and crayfish densities was undertaken from 1992 to 2003. Paua, sea urchins and cat's eyes were monitored in 1992 and 1999 and Cook's turban shells in 1993 and 1999. Data collected so far shows that the average size and numbers of blue cod and butterfish inside the marine reserve have increased while outside the reserve their numbers and average size have decreased. Blue moki and tarakihi numbers were found to be slightly higher inside the reserve and more approachable. Red crayfish are considerably more common in the reserve and the average size is much larger than those from outside. Shellfish and sea urchin numbers varied both inside and outside the reserve and the results were inconclusive. Research on fish and invertebrate species continues at sites inside and outside the marine reserve.

Boulder Reefs

Boulders and rocks dominate the southern end and the eastern and western sides of Long Island. Flapjack seaweeds grow in the shallows, often surrounded by small spotties. Tube worms cover much of the rocks and extend red filaments to trap plankton. Cat's eyes and Cook's turbans are the dominant grazing seashells. Large black-foot paua live under the boulders just below low tide and smaller yellow-foot paua appear at depths of around 5 m. The patchy seaweed disappears at around 8 m. Blue cod cruise in and check out visiting divers but are not as common or inquisitive as those at the northern end of the reserve.

Patches of common anemones cover the rocks and jewel anemones replace them on steep walls and reef underhangs. Sea tulips (sea squirts on long stalks) begin to appear from 5 m, with both white and magenta varieties.

Northern reefs

The reefs between the northern tip of Long Island and the Kokomohua Islands have the most prolific marine life. Patches of red, green and brown seaweeds cover the rocks and schools of spotties swim among the fronds. Banded wrasse and leatherjackets are present in large numbers and small schools of tarakihi swim not far from the reef. Blue cod are prolific and very large. On any dive a dozen blue cod will move in and follow you around. The more aggressive individuals will peer into a dive mask and on seeing its reflection, will attempt to chase its mirrored image away, resulting in a collision with the diver.

Grazing cat's eyes and Cook's turban shells sit on the rocks and several species of whelk crawl among the rocks and seaweeds. Patterned kelpfish sit almost invisible in the shelter of the seaweed forests. Sea urchins congregate on the shallow rocks, grazing on seaweeds. Between the small seaweeds very large green-lipped mussels (some exceeding 18 cm in length) form tight clumps. Hermit crabs occupy many of the small seashells and scurry around, stopping every now and then to peer out from under the rim of their borrowed shells. Sea cucumbers live on the sandy patches between the rocks.

Red crayfish live under the boulders at depths of 5 m and deeper, with some very large animals weighing in at over 3 kg. Their numbers increase around the Kokomohua Islands with groups of six or more animals in some holes.

Large individual blue moki, tarakihi and schools of butterfly perch appear down the reef walls. The seaweeds give way to patches of sponges and ascidians. Sea tulips on long stalks protrude from the reef walls. White bonsai-like hydroids grow to 30 cm high and are grazed on by pink and white Jason nudibranchs.

At around 14 m the rock changes to sand and gravel with more sea cucumbers, some scallops and horse mussels. Leatherjackets, sea perch and blue cod fossick over the sand and some snapper have been reported, albeit in small numbers.

Marine mammals

Bottlenose dolphins often move through the outer Marlborough Sounds in pods of around 15 animals. Less often common dolphins, identifiable by their grey-and-cream colouring, and the acrobatic grey-and-black dusky dolphins visit the reserve. Hector's dolphins are resident in the Sound and small pods visit the reserve to feed. Orca (killer whales) visit infrequently in pods of around eight to 10. They are known to prey on stingrays and possibly feed on dolphins and fur seals, the latter being regular visitors in winter when they occasionally haul out on the rocks.

Coastal flora and fauna

As Long Island has no introduced predators (except the Polynesian rat), it supports a wide range of seabirds and forest birds. It is a scenic reserve and the bush has rapidly regenerated since reverting from farmland in the 1930s. Little spotted kiwi have been released on the island and are thriving. Australasian gannets, probably from the colony on Farewell Spit, regularly feed around the island. Other birds often seen are little blue penguins, king and pied shags, white-fronted terns and shearwaters.

Activities

Snorkelling

Because of the distance from Picton the only snorkelling is from a boat. The reefs right around the island and rocks are worth exploring but strong currents pass close to the northern end of the island.

Scuba-diving

The best diving is around the northern end of the island and the Kokomohua Rocks. Be aware of strong currents offshore and changing wind conditions.

Kayaking

The Marlborough Sounds offer good protected kayaking. The nearest setting off point is Ship Cove for the 5–6 km paddle. Be aware of the exposed location and that sea conditions can change quickly. While Long Island is a considerable distance from Picton it is an interesting paddle. Landing on the island is permitted with the easiest site at the boulder bank on the western side.

Walking

There are no formed tracks and foot access around the shore is difficult.

POHATU (FLEA BAY) MARINE RESERVE

This tiny bay on Banks Peninsula is the only marine reserve on the east coast of the South Island. Yellow-eyed and white-flippered penguins feed in the waters of the reserve and nest on the shore around the bay. Hector's dolphins and New Zealand fur seals are regular visitors.

Created: 1999

Size: 215 ha

Boundaries: The south-western boundary at Ounu-hau Point and the north-eastern boundary at Redcliffe Point are marked by single yellow triangular markers on the cliff tops and extend seaward 506 m and 589 m respectively.

Getting there: From Christchurch, take SH 75 to Akaroa (60–90 minutes' drive). The Lighthouse Road, accessible by four-wheel-drive vehicle only, climbs steeply above Akaroa Harbour and descends to the Flea Bay turn-off at the south-eastern tip of Banks Peninsula. The nearest boat ramp is at Akaroa Harbour. Flea Bay has a launching area for small dinghies and kayaks.

Best time to visit: Low tide is the best time to see the marine life on the rocky shore. Snorkelling and scuba-diving are best at high tide when there is no swell and no southerly wind. White-flippered and yellow-eyed penguins are best seen as they come ashore at dusk to return to their nesting sites.

Activities: Snorkelling, scuba-diving, kayaking, swimming, walking.

Facilities: None.

Rules: Marine Reserve regulations apply (see page 12). Contact the Department of Conservation's (DoC) Christchurch office 03 379 9758. No domestic animals are allowed in the marine reserve.

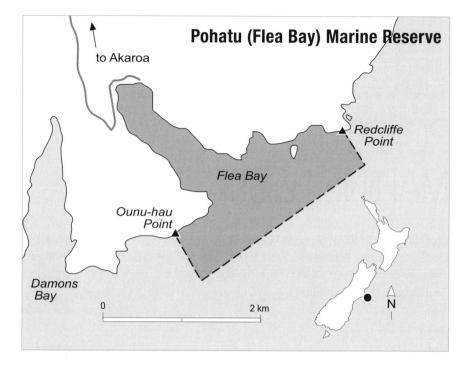

History

Flea Bay, with its abundant fish, shellfish and seabirds, was a rich food source for Maori in pre-European times. Dyke Head, on the south side of the bay, was the site of a pa known as Paekororo. Pohatu, the Maori name for the bay, means 'stone'.

Conservationists first proposed the creation of a marine reserve in Akaroa Harbour in the early 1990s, but local fishers opposed the idea. In 1996 the Canterbury Recreational Marine Fishers' Association and the Akaroa Harbour Recreational Fishing Club made an alternative application for a marine reserve at Flea Bay. Kai Tahu and Te Runaka o Koukourarata, local Maori, supported the protection of Pohatu as a marine reserve to help conserve seafood stocks around Banks Peninsula.

Topography

Banks Peninsula is the remains of an ancient volcano from the Miocene era and was once separated from the South Island. The various bays around the peninsula are the result of massive landslides breaching landlocked craters. Flea Bay on the southeastern exposed coast is surrounded by steep hills several hundred metres high. The bay is affected by cold water from the Southland Current, which tends to bring clearer water into Flea Bay than other areas around Banks Peninsula.

Marine life

With set nets banned, the marine life around the bay has increased noticeably and over 40 species of fish have been recorded.

Rock platform

At the head of the bay is a sand and gravel beach, which makes easy access for snorkelling. Rock pools in the reef platforms surrounding the bay are accessible at low tide. Limpets and chitons sit below little black mussels, barnacles and periwinkles, which dot the high tide line. The pools have thick growths of brown, red and green seaweeds. Crabs, sea urchins, whelks and cat's eye shells cling to the rocks at the edges of pools. Several species of triplefin dart across the pool bottoms, and clingfish and small paua hold on to the undersides of rocks.

At the exposed entrance to the bay, giant bull kelp swirls around the rocks. Colourful sponges and sea tulips (stalked sea squirts) sit among the kelp holdfasts. Patches of blue mussels and, near the low tide mark, larger green-lipped mussels cling to the rocks. After southerly storms and heavy rain the water in the bay is often discoloured by silt washed down from the surrounding hills.

Sheltered reefs

At around 4 m the sandy beach gives way to a mix of rocky reefs, where sea horses hide among dense forests of brown, green and red seaweeds. Sea urchins, brown biscuit stars and spiny sea stars also live among the seaweeds. Large black-foot paua graze on the rocks where sea tulips and sponges mix with other invertebrate life. Long strands of giant bladder kelp float on the surface, aided

RESEARCH

Initial studies of Pohatu Marine Reserve's marine life were carried out in 2000 and 2002. Studies concentrated on fish, paua and rock lobster numbers. No significant differences between the reserve sites and control sites were recorded, but a good baseline has been established for future studies.

Surveys of the white-flippered penguin population at Flea Bay show an increase from 717 pairs in 2000–2001 to 893 pairs in 2005.

by their air-filled bladders.

Spotties, blue cod, banded wrasse and butterfish are the most common fishes among the seaweeds and out into the deeper zones. Crayfish numbers have increased, with nests of animals scattered around the deep cracks and under the boulders. Reports of large crayfish in excess of 4 kg have been noted but most are much smaller.

Deeper reefs

At 10 m and below, giant bladder kelps are generally absent, and the rocks have a coating of red coralline algae. At 12 m the reefs change to a mix of rocks and sand gutters that drop to over 20 m at the mouth of the bay. Tarakihi and trumpeter are common, but blue cod are the dominant fish. Blue cod will approach divers closely, often peering at their reflections in divers' masks. Other fish species include scarlet and girdled wrasse, leatherjackets and the occasional juvenile grouper or hapuku. Blue moki numbers have increased but could not yet be classed as common in the bay.

Little southern pigfish, with their 'sad' eyes and extended noses, hide under pieces of dislodged seaweed. Sponges, ascidians and sea tulips line the rocks and sea and red-banded perch sit among the invertebrate life. Large octopus are not uncommon, and are the favoured food of New Zealand fur seals.

Marine mammals

Hector's dolphins are regular visitors over summer when they feed on small fish in the bay. They are New Zealand's smallest dolphin at 1.5 m and the Banks Peninsula area is one of the best places to see them. New Zealand fur seals also visit with groups of young males hauling out on the rocks around the bay. No breeding colony has been established.

Coastal flora and fauna

Replanting and pest eradication around Flea Bay has increased suitable nesting habitats for white-flippered and yellow-eyed penguins. Both species feed in the surrounding waters during the day and come ashore at dusk to return to their nesting sites amongst the flax and coastal shrubs. White-flippered penguins are the local variant of little blue penguins. Nest boxes have been built for them in addition to natural sites created by replanting. More than 800 pairs of white-flippered penguins currently nest around the bay. Banks Peninsula is the northernmost breeding site for the rarer yellow-eyed penguins. The rocky cliffs and islets near the bay entrance are important nesting sites for spotted shags and oceanic seabirds such as fairy prions and terns.

Activities

Snorkelling

Flea Bay is usually sheltered and offers good snorkelling with easy access from the beach. In summer Hector's dolphins move into the bay and will interact with snorkellers. In winter young male New Zealand fur seals may zoom around and check out snorkellers along the rocky edge of Flea Bay.

Scuba-diving

The shallow reefs of Flea Bay can be easily accessed from shore. The deeper reefs are best dived from a boat or kayak. Although the dolphins normally avoid divers due to the bubbles, they make fleeting passes which can raise adrenaline levels if you're not expecting them. New Zealand fur seals are more likely to check out underwater visitors.

Kayaking

Launch your kayak from Akaroa Harbour and follow the coastline to Flea Bay, a trip of around 15 km, or drive over the hill and launch at Flea Bay. Sea conditions between Akaroa and Flea Bay can be extreme in onshore conditions and heavy swells.

Walking
Banks Peninsula Walk (2–4 days)

This well-known private walk covers 35 km of Banks Peninsula and includes Flea Bay. Transport is included from Akaroa. The walk is across farmland, but follows the cliffs around the coast from Long Bay to Flea Bay with spectacular views of the coast. It passes through some native bush near waterfalls that cascade down steep gullies. The track is open from October 1 to April 30.

Penguin walks (1–2 hours)

Guided walks to see the penguins coming ashore at dusk on private farmland are available with a local operator but must be booked in advance. Visitors are advised not to disturb the birds that move onto the rocky reefs then into the surrounding bushes. Their calls after dusk are a highlight of a visit to the bay.

FIORDLAND MARINE RESERVES

Fiordland, the mainland's most remote and least inhabited region, is the only region where a representative network of marine reserves has been created. Two marine reserves were established in 1993 and a further eight in 2005 as part of the Guardians of Fiordland strategy (see History, opposite). The extremely high rainfall carries plant material into the fiords, discolouring the water. A layer of fresh water sits on top of the salt water, further cutting light levels below. This allows black and red corals and other deep-water organisms to live at abnormally shallow depths. This effect, known as deep water emergence, is common across almost all of Fiordland's marine reserves. Because of the regional focus and similarity in marine life, the authors have addressed Fiordland in a single introductory chapter followed by sub-entries on each reserve.

Getting there: Milford Sound is the only fiord with direct road access. Doubtful Sound is accessed by boat across Lake Manapouri, then by bus across the Wilmot Pass to Deep Cove. The other fiords are accessed by boat or helicopter.

Best time to visit: From early summer to late May there is generally less rain and the air temperature is warmer.

Activities: Scuba-diving, kayaking, walking, tramping.

Rules: Marine reserve regulations apply (see page 12). Contact the Department of Conservation's (DoC) Te Anau office on 03 249 7921. Crayfish, caught outside marine reserves in Fiordland, can be stored for up to two months in some designated areas. Anchoring is restricted in some areas due to the susceptibility of the marine life to anchor damage. Some special areas, known as 'china shops', have additional restrictions. Visitors are advised to bring GPS and to source coordinates from DoC's Te Anau office.

History

Explorer James Cook sailed along the Fiordland coast in 1770, but did not enter any fiords. Doubtful Sound got its name when Cook expressed doubts that there was enough wind to sail his ship, *Endeavour*, into the sound and safely back out. Cook returned in 1773 on *Resolution*, anchoring next to Astronomer Point in Pickersgill Harbour for five weeks. The stumps of some trees cleared by Cook's crew are still visible today. He noted that there were several Maori families living on Indian Island.

In 1791 George Vancouver, who had earlier sailed with Cook, returned to Dusky Sound in command of *Discovery* and *Chatham*. He explored Breaksea Sound and named Vancouver and Broughton Arms.

The first sealers arrived on the Enderby ship *Britannia* in 1792. They began sealing at Anchor Island in Dusky Sound and built New Zealand's first house and first ship.

The Spaniards were the first Europeans to explore Doubtful Sound in 1793. They anchored off Febrero Point and explored the sound by longboat. Fiordland has the only Spanish names on New Zealand charts: Nee Island, Point Quintano, Point Espinosa, Bauza Island and Malaspina Reach are some of the names given by the Spanish that still stand today.

More sealers arrived in 1795 and one of their ships, *Endeavour*, not to be confused with Cook's ship of the same name, became New Zealand's first recorded shipwreck. The crew transferred gear and fittings from the shipwrecked *Endeavour* to the ship built earlier by the sealers, and named it *Providence*. By the 1820s the seal population was decimated and southern right whales became the main focus of whalers as they came into the fiords in winter to calve. A shore whaling station was set up in Preservation Inlet.

Most of Fiordland became a park in 1904, changing to a national park in 1952, and is today New Zealand's largest national park and part of the Te Wahipounamu World Heritage Area. Fiordland's underwater world was only discovered by commercial divers who went below the green surface layer while working on the Manapouri Power Station and tailrace tunnel in the 1970s.

The two original Fiordland marine reserves, Piopiotahi in Milford Sound and Te Awaatu Channel in Doubtful Sound, were proposed by the New Zealand Federation of Commercial Fishermen. But it was quickly evident that granting full protection to these two areas alone was not enough to protect all of Fiordland's habitats.

The concerns of various user and interest groups over the pressures facing Fiordland's unique environment led to the establishment of a forum to discuss the issues. A group calling themselves the Guardians of Fiordland's Fisheries and Marine Environment Inc. (GOFF), and consisting of recreational and commercial fishers, charter boat operators, and representatives of government and local body agencies, local Maori, and community, environmental and marine science groups formulated a marine strategy that would see a wide range of representative habitats over the whole region protected. Eight of these areas were identified as needing complete no-take protection provided by full marine reserve status. In addition, other areas, labelled 'china shops' were identified as significant enough to warrant an extra level of protection above marine reserve status. The strategy also looked at fish and other marine resources with a view to sustaining populations for future generations of New Zealanders. Fish and shellfish quotas were adjusted across all the inner fiords with the taking of some species banned completely in some areas until a review is undertaken. For the complete package to be accepted a change in New Zealand law was needed. The Fiordland (Te Moana o Atawhenua) Marine Management Bill was passed in 2005. By making the proposal reviewable, any future changes in species stocks or discoveries of areas of significance can be addressed.

It is to the credit of this group of far-sighted people that the future of Fiordland's marine environment is guaranteed.

Topography

Fiordland's granite rocks are some of New Zealand's oldest, formed over 500 million years ago. There are 14 main fiords extending over more than 200 km of coastline. Each fiord was created by ancient glaciers, which formed U-shaped valleys. The moraines – walls of rock and rubble pushed down by the glaciers – form a barrier at varying depths inside the entrance of each fiord. The fiord walls drop steeply into a basin that has gradually filled with sediments. At the head of the fiords there are some tidal flats, usually washed by streams.

Marine life

Fiordland has one of the highest rainfalls in the world, receiving up to 11 m per year. Run-off from the mountains creates a permanent freshwater layer that sits on the surface of the denser salt water. The freshwater layer varies in depth from 5 cm to over 10 m and tannins washed out of the vegetation stain it the colour of weakly brewed tea. This cuts down the amount of light entering the sea water and animals normally only found below 30 m live as shallow as 10 m. Although there is life in the mud at 400 m in the deepest parts of the fiords, most species are limited to the top 40 m. This colonisation in shallower water by deep water organisms is known as deep water emergence.

As fresh water flows out of the fiords, a layer of salt water flows in carrying larval open-water species into the fiords in a process known as estuarine circulation. Nearly 200 species of fish have been recorded in Fiordland and the range of seaweeds and invertebrates is even greater.

Shallow outer fiords

Near the sea, giant bull kelps dominate the surface. As most of Fiordland's marine reserves are away from the fiord entrances, bull kelp is minimal. Strap kelp and common kelp form quite dense beds below the intertidal patches of flapjack weed. In these areas the freshwater layer is minimal. Blue and ribbed mussels are the most obvious marine life here.

Large numbers of common triplefin and shrimps dart across the mussel beds and in summer luxuriant growths of sea lettuce cover the zone. Inflated cushion stars sit with their bodies raised and drop on any small fish or shrimp that ventures underneath.

Kelp-coloured sea spiders, up to 7 cm across, live on the kelp fronds. Orange sea anemones, around 2 cm across with turquoise centres, cling to the kelp fronds. Giant bladder kelps that commonly form dense forests around southern New Zealand are represented by the occasional individual plant.

Butterfish, marblefish and banded wrasse swim among the kelp, while schools of oblique swimming blennies swarm above.

Shallow inner fiords

Here the freshwater layer is deeper and species tolerant to it are limited. Sea lettuce is the dominant seaweed along with some Neptune's necklace. Mussels are mainly of the blue and ribbed varieties. In the freshwater zone they are safe from predators, but below in the salt water, spiny sea stars predate them. Common sea urchins, some reaching 30 cm across, graze on the fine seaweeds. Further into the fiords the freshwater layer thickens, reducing light levels and habitats for seaweeds. Telescopefish school in thousands, feeding in the freshwater zone and the top of the salt water. The most prolific fish is the common triplefin, occasionally found in densities of more than 100 per square metre.

Shallow shelves, filled with sand, at 3–5 m support turret shells and bivalves, including fan scallops. The large white feeding spirals of tube worms vanish into their calcareous tubes if any threat to them appears. Larger shelves sometimes contain fallen trees which have their own communities living on them. Common and white sea urchins and several species of sea snail graze on the trees but it's not known whether they feed off the tree or the algae growing on it. Teredo molluscs bore into the trees, helping break them down so that the trees usually disappear within two years.

Large horse mussels protrude from the cliffs, often with only their bases in the sand. Empty blue mussel shells, predated by sea stars, fall here and tube anemones sit among them. Sea cucumbers suck in the sand, remove the bacteria from it, then pass the clean sand out their other end. Strange red-and-white sea cucumbers form dense colonies in some places, extending their tentacles to extract plankton.

Deep walls

Smaller red seaweeds, including coralline algae, cover many of the steep walls below the shelves. Sparse common kelp grows on the less steep walls and green sea rimu and codium seaweeds fill the areas around the common kelp holdfasts down to 10 m. Numerous sea stars and white and pink sea urchins graze on the seaweeds. Fish are prolific with 12 species of triplefin occurring from the shallows, right down the walls. Yellow-black triplefin is the most often seen.

The most common sea slugs are 15 cm warty Wellington nudibranchs and 50 mm gold-lined nudibranchs. Pink-and-white Jason nudibranchs feed on hydroids. At first glance large black shield shells could be mistaken for nudibranchs, but are more closely related to paua as they still have a large shell, often hidden in the folds of the body.

On steep walls where there are no shallow ledges, trees brought down by landslips drop into the depths, taking with them life from the walls. Two fish not common near the entrances, spotties and girdled wrasse, become more common further into the fiords.

Black coral colonies in Fiordland number more than 7 million and the largest

RESEARCH

Early research on the marine species of Fiordland was carried out by dredging, often as part of charting and geological surveys. With the advent of scuba-diving and better research methods, the real value of Fiordland's marine life was realised. Early surveys were mainly fisheries related and some scallop surveys conducted in the 1970s.

Current surveys on a multitude of marine life are being carried out by the National Institute of Water and Atmospheric Research (NIWA), the University of Otago, the University of Canterbury, the Museum of New Zealand and the Department of Conservation (DoC). Some overseas research institutions have joined with New Zealand scientists on specific projects in recent years.

are between three and four centuries old. Below 10 m depth large black coral colonies extend from the walls and create homes for snake stars, which wind themselves tightly around the branches. The snake stars provide a useful service, removing silt from the black corals' polyps. During years of less than average rainfall many shallower black corals die off because of increased light levels. Some black corals are taken over by yellow zoanthids and sponges. Once the rainfall levels return to normal and light levels decrease, a number of the black corals recover.

Colonial sea squirts, which look like sponges, are quite common around the bases of black coral colonies. Lace corals form dense colonies on the steep walls. Gastropod seashells, like the southern tiger shell, graze at the base of black corals and large tiger shells usually live near sponges on the walls. Cook's turban shells, trumpet shells, and circular saw shells are common.

Hermit crabs are prolific and easily identified by their rapid movements across the rocks. Some hermit crab species have a favourite shell home, one preferring only circular saw shells, for instance. Numerous species of sea stars, including several spectacular red species, feed on the molluscs, brachiopods and worms.

Schools of butterfly perch live among the black corals along with leatherjackets, banded, scarlet and girdled wrasse. Blue cod and sea perch, also known as Jock Stewarts, sit and watch. Spotted dog sharks or rig, around 1 m long, dart away if a diver comes too close. Seven-gilled sharks feed in the upper reaches and blue sharks are more common at the outer ends of the fiords. Banded and splendid perch are often seen in the cracks and caves below 20 m. In the southern fiords wavy lined perch, usually found below 100 m, live at a depth of less than 20 m. Red cod and very large conger eels, occasionally more than 2 m long, live in the holes, along with large numbers of red crayfish.

The underhangs are covered with invertebrate growth, with red hydrocoral colonies, 30–40 cm across, the most spectacular. Channels and pinnacles with higher water current have the most impressive numbers of red corals, octocorals and solitary corals.

In some places the walls are completely covered in bivalve-like brachiopods. These ancient two-shelled animals date back more than 500 million years and are often mistaken for seashells. Other names for them are lamp shells and ladies' toenails.

Sponges, while quite common outside the fiords, are limited inside. The situation varies from fiord to fiord, with most having around a dozen species. Fifty species of sponge have been recorded in Fiordland, but only grey cup sponges are found in large numbers.

Deep sandy areas
Large numbers of scallops and horse mussels live on the sand at the bases of steep cliffs between 20 and 40 m. Queen scallops up to 15 cm across and a smaller species, grooved fan shells up to 6 cm across, appear in reds and browns. Similar sized but drab-coloured southern file shells live buried in the sand.

Preservation Inlet and Te Awaatu Channel are two areas that support sea pens, which rely on the current to bring food particles to their feather-like polyps. They stand up to 30 cm high. Five species have been identified in Fiordland.

Marine mammals
Bottlenose dolphins regularly visit many of the fiords and are resident in Doubtful Sound. They interact with kayakers and divers and ride boat bows. New Zealand fur seal numbers are now recovering in colonies near the fiord entrances. They often swim up the fiords to feed and play. Leopard seals and New Zealand sea lions are occasional visitors. Southern right whales and sperm whales were once regular visitors to the fiords, but now their numbers have been reduced by whaling they are rarely seen.

Coastal flora and fauna
The steep hillsides surrounding the fiords are typical of southern podocarp forests. More than 700 plant species have been identified in Fiordland and of these 24 are endemic to the area. Introduced pests such as deer, possums, rats and stoats have severely affected many of the bird populations, but attempts to reduce the pests have successfully increased some bird populations.

Fiordland crested penguins, now among the world's rarest penguins due to predation, and little blue penguins live and feed around the fiords.

Activities

Scuba-diving
Most scuba-divers in Fiordland dive off charter boats and skippers have good knowledge of dive sites. Many dive sites are alongside steep walls and divers must have good buoyancy skills. Special care must be taken with fins and loose dive gear so you don't damage the fragile black and red corals, which may have taken hundreds of years to grow.

Kayaking
Milford and Doubtful Sounds are the usual setting-off points for kayak trips. Longer guided kayak tours explore Doubtful and Dusky Sounds.

Walking

Although there are a large number of walking tracks in Fiordland only a few reach the fiords.

Milford Track (4 days – one way)
Fiordland's best-known walking track can be done as part of a guided party or as an individual but a prebooked accommodation pass is needed from DoC. Transport is required to the start of the track at the northern tip of Lake Te Anau. The track travels through deep rainforested valleys, past lakes and waterfalls and beneath towering peaks. It finishes 53 km later at Milford Sound.

Milford Lookout (10 mins return)
This short track begins behind the hotel where steep stairs climb through native forest to a lookout that gives views over Milford Sound.

Bowen Falls (30 mins return)
An easy walk on a good track that begins at Freshwater Basin and leads to the spectacular Bowen Falls. Take a raincoat, even on a fine-looking day, if you want to stay dry.

Astronomer's Point, Dusky Sound (20 mins return)
This is the point where explorer James Cook anchored in 1773. You can still see the stumps of the large trees that were cut down by his crew. A boardwalk leads through native bush past the site of the first brewery in New Zealand. There is also a side track to Lake Forster.

PIOPIOTAHI MARINE RESERVE

Created: 1993
Size: 690 ha
Getting there: From Te Anau take SH 94, through the Homer Tunnel, to Milford Sound.
Facilities: Toilets, accommodation, boat ramp and boat tours. Milford Sound Underwater Observatory in Harrison Cove is used for tourism, education and scientific research.
Boundaries: The seaward boundary runs approximately along the centre of Milford Sound and is marked by a series of red port markers. The north-western point is at Dale Point and the southern boundary is near Deepwater Basin where it extends to the land.
Rules: Marine reserve regulations apply (see page 12).

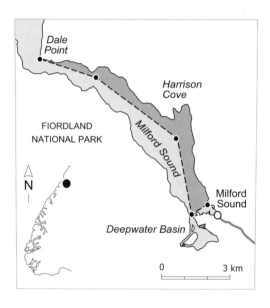

History

Piopiotahi means 'one native thrush', a ground-feeding bird now believed to be extinct. Captain John Grono was the first recorded explorer to sail into the sound, which he named Milford Haven after his place of birth in Wales. The area was opened up with the building of the Milford Road and Homer Tunnel. The creation of a marine reserve was first proposed by the New Zealand Federation of Commercial Fishermen.

Marine life

The steep hills above the reserve plunge to depths of 300 m in the sound. Black coral colonies here are not as large as those in some southern fiords. It is possible that damage from tree avalanches limits their size. There are many thickets of red corals on the wall underhangs, some as deep as 60 m. At several sites there are spiny sea dragons, a relative of the sea horse. Most of the marine life found in Milford Sound is typical of Fiordland.

TE HAPUA (SUTHERLAND SOUND) MARINE RESERVE

Created: 2005
Size: 449 ha
Boundaries: The reserve encompasses the entire inner Sutherland Sound.
Rules: Marine reserve regulations apply (see page 12).

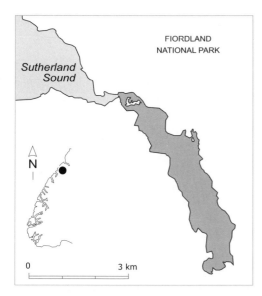

History

Sutherland Sound was named after Donald Sutherland, the first European to live in Milford Sound. He explored much of Fiordland and discovered the Sutherland Falls.

Marine life

Sutherland Sound has a very shallow entrance and differs from all the other fiords in that it lacks an inflow of salt water. The rocky moraine near the entrance is exposed at low tide and the sound itself is a shallow, muddy estuary with accumulated plant material. Dogfish, stargazers, flounder and red decorator crabs are all common here. Because of its shallow nature it has not been affected by human interference and is relatively pristine.

HAWEA (CLIO ROCKS) MARINE RESERVE

Created: 2005
Size: 411 ha
Boundaries: Bligh Sound, between Turn Round Point and Clio Rocks.
Rules: Marine reserve regulations apply (see page 12). Anchoring is prohibited off Clio Rocks.

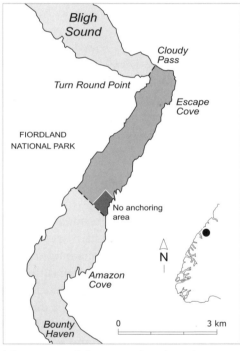

History

In 1809 Captain Grono explored Fiordland on his ship *Governor Bligh*. He named Bligh Sound after the former governor of New South Wales, Australia. Clio Rocks were named after HMS *Clio*, which struck an uncharted rock in 1871.

Marine life

Hawea Reserve runs between two of the areas designated as 'china shops'. At Turn Round Point there are spectacular black coral colonies and intense invertebrate growths. Sponge life is some of the most abundant discovered in Fiordland. The waters of Bligh Sound are clearer than most, due to less run-off. At the opposite end, around Clio Rocks, steep pinnacles rise from 200 m to near the surface in the centre of the sound allowing a settling point for a wide variety of sessile marine life. Black, red and pink coral colonies are prolific on these pinnacles, along with a great diversity of other encrusting life.

KAHUKURA (GOLD ARM) MARINE RESERVE

Created: 2005
Size: 464 ha
Boundaries: The entire Gold Arm, Charles Sound.
Rules: Marine reserve regulations apply (page 12). Anchoring is prohibited in part of the marine reserve.

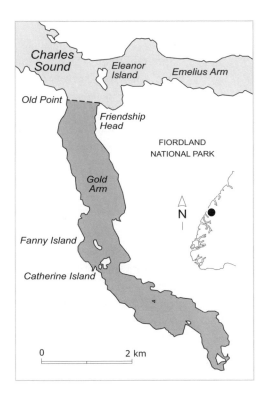

Marine life

This marine reserve encompasses the 'china shop' identified by GOFF as containing a complete representation of the habitats within Charles Sound. The marine reserve includes river mouths, estuarine areas, islands and tidal rocks. Black and red corals live very close to the surface and can be seen from boats on a calm day. Schools of most fish species typical of Fiordland are found in this area. Red crayfish numbers are higher than in some of the other fiords.

KUTU PARERA (GAER ARM) MARINE RESERVE

Created: 2005
Size: 433 ha
Boundaries: The entire Gaer Arm, Bradshaw Sound.
Rules: Marine reserve regulations apply (see page 12). Anchoring is prohibited in part of the marine reserve.

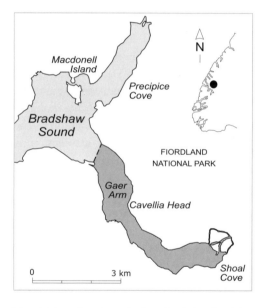

History

Captain John Stokes on HMS *Acheron* surveyed the New Zealand coast between 1850 and 1851. He named Bradshaw Sound after one of his midshipmen.

Marine life

The upper estuary has large cockle beds and some sea pens in deeper water. Soft corals are abundant on the walls and there are excellent fish populations. The 'china shop' at the head of the fiord has a variety of fish species including opalfish, grouper and tarakihi. Colonies of both black and red corals are plentiful and red crayfish populations are reasonable.

TE AWAATU CHANNEL (THE GUT) MARINE RESERVE

Created: 1993

Size: 93 ha

Boundaries: The north-western boundary is marked by white triangular markers at Grono Bay on Secretary Island and Bauza Island. The southern boundary has two triangular markers at Marcaciones Point. The boundary extends 1 nautical mile south-east of Marcaciones Point at which point two triangular lead markers at Blanket Bay on Secretary Island line up.

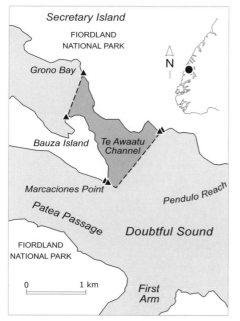

History

Te Awaatu is also known as Te Awa-o-Tu, which translates as 'the channel of Tu'. Maori myths and legends tell of Tu-Te-Raki-Whanoa, who carved out the fiords and lakes with his giant digging stick, or ko. According to legend, he placed his feet on the two largest islands in Fiordland, Secretary Island (Ka-Tu-Waewae-o-Tu) at the entrance to Doubtful Sound and Resolution Island (Mauikatau) at the entrance to Dusky Sound.

The first Europeans to enter Doubtful Sound were Spaniards under the command of Italian Alessandro Malaspina, in 1793. Don Felipe Bauza explored the sound by longboat and landed at the eastern tip of the island, which is named after him. He also landed at a little bay on Secretary Island named Grono Bay.

Marine life

Te Awaatu is the smallest marine reserve in New Zealand, yet it is home to a wide variety of fish and invertebrates. The freshwater layer is usually around 3 m deep.

The maximum depth is around 45 m and the channel is often affected by fast-flowing currents. On the sand are colonies of sea pens. These invertebrates, which look like Victorian-era quill pens, stand up 20–25 cm from the sand and filter passing plankton.

On the fiord walls and underhangs are red and black corals, hydroids and a profusion of other invertebrates. Snake stars, in a variety of colours, wind tightly around the black coral branches and disentangle themselves at night to feed. Large snake stars although similar in size and shape, feed on the sandy shelves on the fiord walls. Schools of fish species, typical to all of Fiordland, are also found here. Strange bivalve-like lamp shells or brachiopods cover much of the fiord walls, often with coatings of coralline algae.

TAIPARI ROA (ELIZABETH ISLAND) MARINE RESERVE

Created: 2005

Size: 613 ha

Boundaries: Malaspina Reach to Deep Cove including around Elizabeth Island, Doubtful Sound.

Rules: Marine reserve regulations apply (see page 12). Anchoring is prohibited in part of the marine reserve.

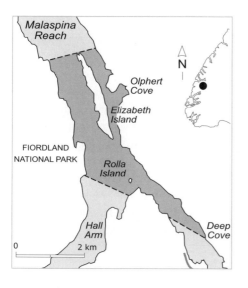

History

Malaspina Reach was named after the Italian captain of the Spanish ships *Descubierta* and *Altrevida* which explored Doubtful Sound in 1793. Captain Stokes of the survey ship HMS *Acheron*, unaware that it had already been named, gave it the name Smith Sound. However, it later reverted to its original name. Captain John Grono named Elizabeth Island after his wife. His ship on the 1821 expedition was also named *Elizabeth*.

Marine life

Five 'china shops' were identified in this area, which is close to the Deep Cove wharf and road access over Wilmot Pass from Lake Manapouri. The freshwater layer is usually deeper due to the outflow from the Manapouri Power Station. At the southern end of Elizabeth Island are excellent red coral colonies, which extend down to the sandy floor. Opposite Hall Arm are colonies of extremely rare bright yellow glass sponges at depths of 30–35 m. Although much of the area is barren, other excellent red coral colonies live on the rock wall underhang south of Lady Alice Falls and inside Rolla Island and Tarawera Rock. Off Brigg Point a white coral colony exists in the trench.

MOANA UTA (WET JACKET ARM) MARINE RESERVE

Created: 2005
Size: 2007 ha
Boundaries: The entire Wet Jacket Arm off Acheron Passage.
Rules: Marine reserve regulations apply (page 12). Anchoring is prohibited at the entrance to Wet Jacket Arm.

History

In 1773 Cook's *Resolution* was becalmed at Detention Cove for a few days and several crew explored Wet Jacket Arm in a longboat. They camped near the headwaters, but woke to strong winds and rain. Bad weather prevented them from returning to their ship, and they spent another night in the bush, soaking wet, hungry, and with no shelter from the extremely stormy weather. On their return to the *Resolution* they named the arm Wet Jacket.

Marine life

This marine reserve was identified as a typical strong current, rock wall habitat and contains examples of all the inside fiord habitats in a small area. The 'china shop' on the sill at the entrance to Wet Jacket Arm is not affected by silt and supports excellent invertebrate communities, including large black coral colonies and bryozoans. Black corals are the densest in Fiordland and large brachiopod beds consist of at least five species.

The head of the arm is a tidal sand flat with cockle and pipi beds where large numbers of estuarine triplefin live. Black corals live near the head of the sound. There are large numbers of blue cod and grouper. Night dives can sometimes attract seven-gill sharks, which feed in the shallows.

Top: *Pohatu Marine Reserve at Flea Bay, on the south-eastern end of Banks Peninsula, is Canterbury's only marine reserve to date.*

Above left: *Little orange anemones hide on kelp fronds at Pohatu.*

Above right: *Yellow-eyed penguins are at the northernmost limit of their range on Banks Peninsula.*

Left: *Snake stars wind tightly around black coral colonies during the day, unwinding at night to feed.*

Below far left: *Red corals, found in shallower waters in Fiordland than elsewhere, continue growing for centuries.*

Below left: *A variety of beautiful worms live on the reef walls in Fiordland.*

Below: *Fiordland's reduced light levels allow black coral colonies to live at unusually shallow depths.*

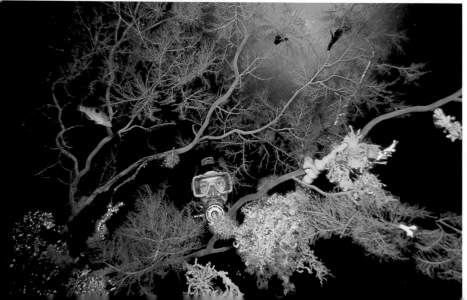

Above: *Sea pens live in the sandy channel of Te Awaatu Marine Reserve.*

Above right: *Tranquil reflections; Fiordland is home to no fewer than 10 marine reserves.*

Below right: *Sea spiders cling to the undersides of kelp fronds.*

Below far right: *Biscuit sea stars, found in New Zealand's southern waters, vary in colour from dull brown to red.*

Above: *Post Office Bay on Ulva Island, a wildlife sanctuary within Te Wharawhara (Ulva Island) Marine Reserve.*

Left: *Large black-foot paua live in the cool, algae-rich waters of Te Wharawhara.*

Above right: *The Auckland Islands are the main breeding grounds for New Zealand sea lions.*

Below right: *Weka feed among washed-up seaweeds along the tideline at Te Wharawhara.*

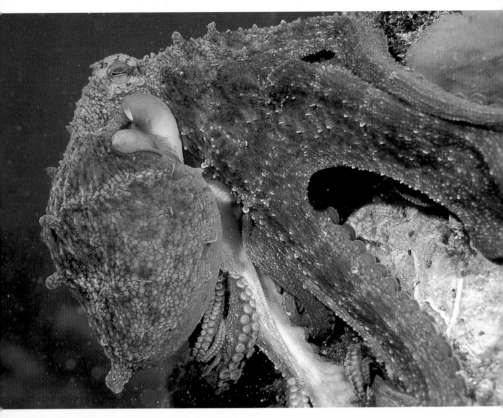

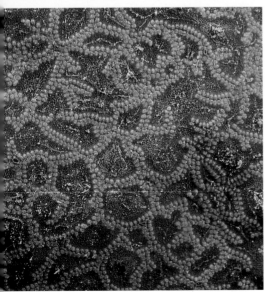

Above: *Large sand octopus, up to 2 m across, are found right around the New Zealand coast.*

Left: *Surface detail of a colonial sea squirt. These invertebrates come in a wide variety of colours, shapes and sizes and can resemble mosses, corals or seaweeds in appearance.*

TAUMOANA (FIVE FINGERS PENINSULA) MARINE RESERVE

Created: 2005
Size: 1466 ha
Boundaries: Inside Five Fingers Peninsula the boundary runs due east to a point just south of Pigeon Island and includes Parrot Island.
Rules: Marine reserve regulations apply (see page 12).

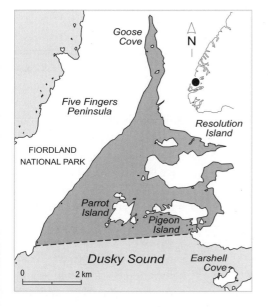

History

The explorer James Cook sailed close to Five Fingers Peninsula on his first voyage to Fiordland in 1770 and named the area after the appearance of the rocks that were, in his words, 'standing up like the four fingers and thum [sic] of a man's hand'. As it was just getting dark he decided not to enter Dusky Sound and continued north.

Facile Harbour inside the marine reserve is the site of New Zealand's first shipwreck, *Endeavour*.

Marine life

Although this area does not contain any 'china shops', it has a good representation of habitats. These include rocky reefs, sandy floor, estuaries and kelp forests. Large scallop beds exist on the sandy areas, along with associated species. Some sea pens have been reported here also. Generally the freshwater layer is less than 3 m and large areas are covered with common kelp and to a lesser extent giant bladder kelps.

Many of the fish species from Fiordland are found in this area and blue cod numbers were found to be exceptional, with some very large specimens.

TE TAPUWAE O HUA (LONG SOUND) MARINE RESERVE

Created: 2005
Size: 3672 ha
Boundaries: The entire area from Long Sound to Useless and Revolver Bays.
Rules: Marine reserve regulations apply (see page 12). Anchoring is prohibited in part of the marine reserve.

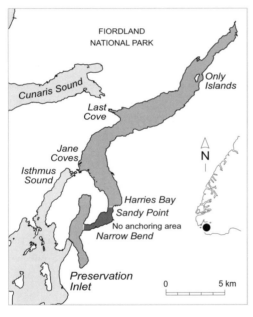

Marine life

The Narrow Bend between Long Sound and Useless Bay was given 'china shop' status due to the abundance of sea pens in shallow water among scallops. It is one of the few places where starburst sea pens have been found. These animals differ from typical sea pens in having branching transparent arms.

Strawberry sea cucumbers can be distinguished from conventional sea cucumbers by their long tentacles, and live just below the freshwater layer. When contracted they resemble strawberries. White sea cucumbers have also been recorded here.

The fiord walls are covered with invertebrate life, including large red coral colonies on the underhangs and white brachiopods on the vertical walls.

Although fish numbers are generally not as high as other areas of Fiordland, southern splendid perch are more common here than anywhere else in Fiordland. Rock wall invertebrate communities are prolific right up towards the head of the sound, with black and red coral colonies, sponges and other invertebrate life.

TE WHARAWHARA (ULVA ISLAND) MARINE RESERVE

Stewart Island/Rakiura's remoteness, clear water and relatively unspoilt marine life make diving and snorkelling here an unforgettable experience. Forests of red, green and brown seaweeds harbour many species of fish including sea horses plus paua and red crayfish.

Created: 2004

Size: 1075 ha

Boundaries: The marine reserve is divided into three sections, centred around Ulva Island. The northern section covers an area from the western point of Native Island around the southern end of the island to the eastern point. The boundary then extends due south for approximately 1 nautical mile, then due west to the north-east end of Ulva Island. It follows the coast to the west to Manawahei Nugget, then to the western point of Native Island. A second, small section covers Sydney Cove on Ulva Island. The boundary extends from Flagstaff Point to Goat Island then to the eastern end of Sydney Cove. The boundary of the larger southern section runs from the western tip of Ulva Island, taking in all the coast to west of The Snuggery, then south-west to West Paua Beach and follows the coast to Trumpeter Point. From there it runs northeast to the western tip of Tamihau Island and the western tip of Ulva Island.

Getting there: Water taxis leave from Halfmoon Bay or Golden Bay for Post Office Bay on Ulva Island. Access to the marine reserve is by boats and kayaks and these can be launched from Halfmoon Bay, Deep Bay, Watercress Bay, Thule Bay and Golden Bay.

Activities: Snorkelling, scuba-diving, swimming, kayaking, walking.

Best time to visit: Summer during periods of calm.

Facilities: Ulva Island has a public toilet and shelter.

Rules: Marine reserve regulations apply (see page 12). No domestic animals are permitted on Ulva Island.

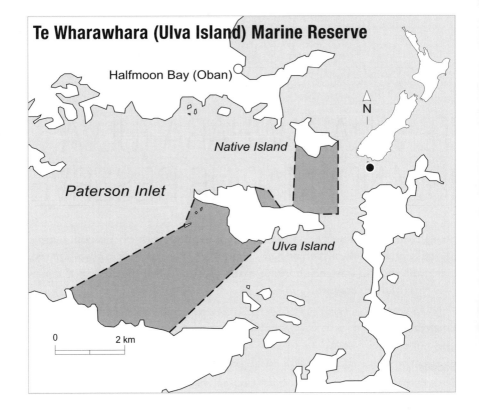

History

Early Maori came to live on Stewart Island around 700 years ago, creating the southernmost Polynesian settlement. Evidence of their middens can be seen around Paterson Inlet, or Te Whaka a Te Wera as it is known by Maori. Explorer James Cook sailed past Stewart Island in 1770 but thought the land was part of the South Island. Sealers arrived in the late 1700s and were followed by whalers in the early 1800s. Charles Traill settled on Ulva Island in 1870 and opened a general store and post office which operated until 1923. The Post Office building still stands today.

The proposal for creating a marine park in Paterson Inlet was lodged by the Ministry of Agriculture and Fisheries (MAF) in 1986. This application was then taken over by the newly formed Department of Conservation (DoC) when it took over the responsibility for marine reserves in 1987. DoC put out a questionnaire in 1989 and found strong support for a marine reserve at Stewart Island, with Paterson Inlet highlighted as a favoured site. A committee representing Maori, commercial and recreational fishers, local residents and tourism operators was established in 1991 to consider all the options. The boundaries of the marine reserve were finalised by this committee, working together with DoC, MAF and the Southland Regional Council. The Te Waka a Te Whera/Paterson Inlet Mataitai Reserve, which surrounds the marine reserve, was established in conjunction with Te Wharawhara (Ulva Island) Marine Reserve.

Topography

Paterson Inlet, on Stewart Island's east coast, is a submerged river valley, flooded around 10,000 years ago when seas rose after the last ice age. The inlet covers 8900 ha and 188 km of coastline, and encompasses 20 small islands, the largest being Ulva. The inlet is almost completely enclosed, opening out to Foveaux Strait. Freshwater and Rakeahua are the main rivers flowing into the inlet, but neither carries much sediment as the land is mainly covered in virgin forest.

The shoreline is granite and diorite broken rock and platforms with an area of crushed schist along the southern shore. Small sandy beaches and tidal estuaries dot the shoreline. Most of the inlet is less than 20 m deep.

Stewart Island lies in the path of the Roaring Forties, the strong westerly winds of the Southern Ocean. The Subtropical Front, a current from the west, brings a milder, less extreme climate than that of neighbouring Southland.

Marine life

Paterson Inlet has a diverse range of underwater habitats. Seven different communities have been identified and most are represented in the marine reserve. It is the only known place in the world where brachiopods, dating back 500 million years, are the dominant marine life.

Tidal reefs
Along the high tide mark are hard granite rocks with patches of barnacle, Neptune's necklace and lower down several species of limpets. In spring green sea lettuce is common. Blue mussels form small clumps near the low tide mark among dense brown seaweeds. Several species of grazing seashells feed on the luxuriant seaweeds in the rock pools. Top shells and small paua are common.

Seaweed forests
The dominant organisms of the underwater world at Paterson Inlet are the seaweeds with around 270 species recorded. Small brown, red and green

seaweeds cover much of the shoreline. Bull kelp grows around the exposed headlands at the outer extremes of the reserve. Giant bladder kelp dominates from where bull kelp stops and creates important habitats and nurseries for marine fish. In the clear water of the inlet bladder kelp grows at depths of 20 m or more. On some sandy areas seaweeds give cover for invertebrate life and surfaces for larval settlement.

Orange-and-green striped anemones live on the kelp fronds and opal top and Cook's turban shells graze on the seaweeds. Colourful southern fan scallops are sometimes engulfed in sponges matching the colours of the shells. When scallops are partially open, their brilliant blue eyes show as pinpoints just inside the shells. Brachiopods – primitive two-shelled, bivalve-like animals – are represented by many species. In some shallow areas, they are the dominant encrusting life, especially in the outer part of the inlet and east of Ulva Island. Brachiopods are also known as lamp shells or ladies' toenails since most are bright red.

Around 56 fish species have been recorded in Paterson Inlet. Amongst the kelp forest are blue cod, banded, girdled and scarlet wrasse, leatherjackets, butterfly perch, blue moki, southern pigfish, butterfish (greenbone), tarakihi and trumpeter. Sea horses and pipefish live among the smaller seaweeds. Biscuit sea stars, up to 10 cm across, are usually dull brown or chocolate colour.

Sea urchins and black and yellow-foot paua live among the sponges, anemones and sea tulips around the kelp holdfasts. Red crayfish and grouper (hapuku) were once very common but numbers decreased before creation of the reserve.

On the sand

At 10–15 m the kelp-covered reefs give way to sand habitats for scallops, oysters, opalfish, ling, skates and electric rays. Red and green seaweeds grow on the sand and help stabilise sand movement. Often little southern pigfish sit camouflaged under loose brown seaweeds. Off the eastern tip of Ulva Island there are strong tidal currents and important habitats for juvenile fish.

Among the red seaweeds on the sand are wandering anemones and large red snake stars. Octopus sit undetected and often feed on scallops. The oysters for which Foveaux Strait is famous are scattered irregularly over the sand patches below 15 m. Large volute shells, with brown lightning patterns, lie buried in the sand, where they hunt for small bivalves.

Marine mammals

Bottlenose dolphins are regular visitors to the inlet. New Zealand fur seals and less often New Zealand sea lions haul out along the inlet. Remember that these marine mammals can be aggressive on land and approaching within 5 m of them is not advisable.

Coastal flora and fauna

The shores of Paterson Inlet and Ulva Island are surrounded by podocarp forest, which includes ancient rimu and totara trees. Scrub covers the more exposed coast. Pines, macrocarpa and beech trees were planted on Ulva Island during the late 1800s.

Ulva, the largest island in Paterson Inlet, has been a wildlife reserve since 1899. It is a predator-free, open sanctuary with many rare birds. Yellow-eyed and little blue penguins nest and feed around the marine reserve. Shags and sooty shearwaters also nest in the area.

Activities

Snorkelling

Post Office Bay on Ulva Island is the easiest snorkelling site to get to. The shallow reefs are a forest of colourful seaweeds and fish. Be aware of strong currents offshore. The beaches on Native Island can be accessed for snorkelling if you arrange to be dropped off and later picked up by water taxi.

Scuba-diving

There are many good dive sites around Ulva Island and Native Island. Be aware of strong currents in the channels. There's always the chance of an encounter with a New Zealand sea lion or New Zealand fur seal. Their open-mouthed threat display can be unnerving, but they veer away at the last minute.

Kayaking

This is one of the best ways to explore the islands and marine reserve in Paterson Inlet. Launch kayaks at Golden Bay. There are many sandy beaches suitable for landing. Foveaux Strait can be extremely rough so venturing outside the inlet is not advised.

Walking

Landing on Ulva Island is usually at the Post Office Bay jetty, and this is the starting point for most of the walks on the island. All tracks are well formed and make for easy walking. Toilets are at Post Office Bay. Remember the island is predator-free. Take your rubbish away with you and don't feed the weka.

Flagstaff Point Lookout (20 mins return)

From Post Office Bay follow the track through native bush to the flagstaff erected by Charles Traill. The flag was raised when the mail boat arrived from Bluff so residents around the inlet could pick up mail at the Post Office. It joins the track to Sydney Cove.

Boulder Beach Walk (1 hr return)
Begin at Post Office Bay and follow the signs through native bush across the island to the southern section of the marine reserve at Boulder Beach. The boulders the beach is named for are now hidden by the forest at the southern end of the beach. Weka often forage amongst the washed-up seaweeds and other native birds can be seen along the track.

Sydney Cove (15 mins return)
From Post Office Bay it is a short walk to Sydney Cove which has a good beach, picnic tables, shelter and fresh water. Native birds, especially bellbirds, are common in the trees and weka are often seen on the beach.

Other walks on Ulva Island take in West End Beach, Roger Bay and The Snuggery, which are outside the marine reserve.

AUCKLAND ISLANDS (MOTU MAHA) MARINE RESERVE

Shipwrecks with rich cargoes and gripping tales of survival have helped to make these remote southern islands famous, but they are also known for their wealth of marine life and are an important breeding site for New Zealand sea lions, albatrosses, petrels and penguins.

Created: 2004

Size: 498,000 ha

Boundaries: The marine reserve boundary extends offshore 12 nautical miles around the islands.

Getting there: The Auckland Islands are 460 km south-southwest of Bluff and only organised seagoing vessels can get there. The boat trip from Bluff takes 30–48 hrs depending on weather conditions.

Best time to visit: Summer, although weather conditions at this latitude can make access to the islands difficult at any time of year.

Activities: Snorkelling, scuba-diving, birdwatching, marine mammal watching.

Facilities: None.

Rules: Marine reserve regulations apply (see page 12). A permit from the Department of Conservation (DoC) is required to land on the islands. Contact DoC's Invercargill office 03 214 4589.

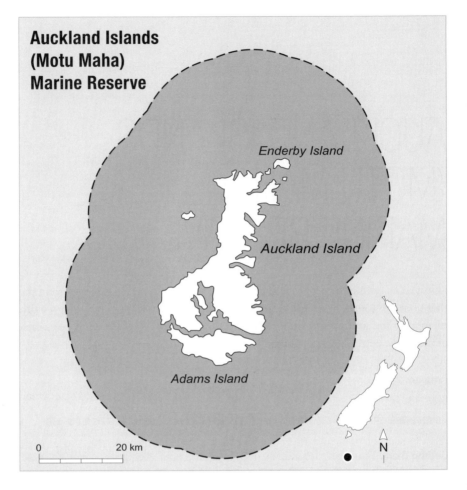

Auckland Islands (Motu Maha) Marine Reserve

History

While there is no record of colonisation of the Auckland Islands by the first Polynesians to settle in New Zealand, Maori oral histories, such as those of Waitaha and Kati Mamoe, record an association with the islands. The oral records of Ngai Tahu recount fishing, bird hunting and other food gathering expeditions to the islands. The islands are also significant to Moriori and to Ngati Mutunga of the Chatham Islands, whose oral histories refer to the two main islands as Motu Maha and Maungahuka respectively.

The first Europeans to discover the Auckland Islands were those on board the whaling ship *Ocean* in 1806. Captain Bristow of the English shipping company, Samuel Enderby and Sons, named the islands after Lord Auckland, previously the first lord of the British Admiralty and Governor General of India. He noted a healthy seal population and returned the following year, claiming the islands for Britain. Pigs were released on Enderby Island as food for castaways.

The seals were killed for their skins and oil, and by 1830 their numbers had been decimated. In 1842 a group of Maori with Moriori slaves came from the Chatham Islands and established a settlement near the entrance to what is now known as Port Ross.

In 1849 Charles Enderby set off from Plymouth in the United Kingdom intending to colonise the Auckland Islands and establish a whaling station. He had been told that the islands had good soil and that once the luxuriant forest was felled, the land would make good pasture. The climate was described as mild, temperate and salubrious. When the three ships arrived at Port Ross they were greeted, to Enderby's surprise, by Maori and Moriori. (He had been told the islands were uninhabited.)

The new settlers soon realised the terrain was inhospitable with peat swamps, tangled scrub and a cold, wet and windy climate. The settlement, known as Hardwicke, was abandoned by 1852. The Maori held out a little longer but they too left the islands in 1856.

Over the next 50 years the islands were the site of many shipwrecks, including *Grafton* and *Invercauld* in 1864, *General Grant* in 1866, *Derry Castle* in 1891 and *Dundonald* in 1907.

Sheep and cattle were released on the islands as food for shipwrecked sailors and mice, cats, goats, rabbits and pigs were introduced. These have since been removed or eradicated.

Adams Island had no introduced pests and was the first to become a wildlife reserve in 1910. The other islands became protected reserves in 1934. In 1986 the islands became a national reserve and now have World Heritage Status as part of the subantarctic islands.

The large number of sea lions dying in squid fishers' nets was a prime motivator for the islands receiving marine reserve status in 2004.

Topography

The Auckland Islands lie in the Southern Ocean, at around 50 degrees south latitude. The two main islands, Auckland and Adams, are the remains of volcanoes aged between two and seven million years. The western sides are lined with very high cliffs, some several hundred metres high, with huge sea caves and frequent rock falls. Streams pour over the cliff edges, changing to waterfalls, which are often forced upwards and scattered by the prevailing winds.

The sheltered eastern sides of Auckland Island slope down into deep bays and long narrow inlets. Port Ross, on the northeast side of the island, is protected by Rose, Ewing and Enderby Islands at its mouth. Enderby Island has a few sandy beaches backed by sand hills.

The igneous rock of Disappointment Island, 6 km to the west, has withstood the extreme prevailing weather conditions that pound the island group's western coast. The water around the islands drops quickly to 100 m and then to 3000 m within the marine reserve boundary.

Marine life

Intertidal zone
Large amounts of seaweed and kelp wash up on the beaches and decompose. Clouds of insects swarm above the kelp and breed among the fronds. Large rata trees provide shade and there is nearly always cloud cover, rain and freshwater run-off.

Western exposed coast
The islands' seaward western side and high cliffs are pounded by mountainous seas for much of the year, minimising marine life. The rocky reefs beneath the cliffs are dominated by mobile boulders. This makes life difficult for encrusting organisms, which can survive only in the few cracks and crevices in the reefs. Few divers, apart from hardy treasure hunters, venture onto this side of the islands.

Eastern sheltered bays
On the eastern side of the islands are sandy bays, separated by rocky headlands with luxuriant kelp growth. The tidal range is around a metre, so there is little by way of an intertidal zone. In the shallows are large patches of green seaweeds and a few grazing seashells. Triplefin are usually the first fish seen. Bull kelp dominates the exposed shallows along with some smaller brown seaweeds. Giant bladder kelp forests thrive in the protected areas from 8 m. New Zealand sea lions can appear and interact with divers at any stage. While juveniles will interact and play, the larger male sea lions will approach with a mouth wide

RESEARCH
Little was known about the underwater life at the Auckland Islands so an initial baseline study was undertaken for DoC in February 2004.

open threat display, but they usually veer away at the last minute.

Dense colonies of blue and ribbed mussels line the reefs. Giant spider crabs, with orange and red shells more than 25 cm wide, are the major predator of the mussels. Huge aggregations of these crabs occur under the kelp forests, probably to mate. Seashell species include the grazing virgin paua, which can reach 70 mm in length, the southern tiger shell and southern opal top shell.

Trumpeter, girdled wrasse and telescope fish are the most commonly seen large fish. Maori chiefs, black cod and Antarctic cod are usually seen resting on rock ledges.

Deep reefs

Schools of blue moki and trumpeter live among the kelp forests which extend down to 30 m. Offshore in deep water are arrow squid, southern blue whiting, scampi and hoki. Great white sharks are known to feed in the shallows close to the islands and most likely prey on seals and sea lions.

Marine mammals

The Auckland Islands are the main breeding ground for endemic New Zealand sea lions, once called Hooker's sea lions. Since the days when they were hunted almost to extinction for their oil and skins, their population has recovered and is now estimated to be up to 15,000. Enderby Island is the main breeding site.

New Zealand fur seals and elephant seals also breed on the islands during summer. Leopard seals visit occasionally to feed on penguins.

Southern right whales were also hunted by whalers but are now starting to make a comeback. During winter they move into Carnley Harbour and Port Ross to breed, with up to 200 animals seen. Orca (killer whales) have also been seen around the islands and probably feed on seals and sea lions.

Coastal flora and fauna

Southern rata trees line the coast and the islands are the nesting place for a wide variety of birds. Fifty-two breeding species have been recorded, including three main species of penguin. Rockhopper penguins arrive at the islands in October to breed and lay their eggs a month later. They leap from the water onto wave-washed rocks and hop with both feet together. Erect-crested penguins are seen in lesser numbers, but breed near the rockhopper colonies. Yellow-eyed penguins live in burrows or under overhanging trunks or roots in the forest.

Northern giant petrels, dark brown and almost the size of an albatross, nest among the tussock and scrub. Other petrels include white-headed, white-chinned, storm and subantarctic diving petrels.

Seventeen species of albatross, including royal, Gibsons, white-capped and light-mantled sooty, can be seen. Terns include Antarctic and white-fronted. Black-backed and red-billed gulls are common. Other seabirds breeding on the islands include cape pigeons, Antarctic prion, fulmar prion, southern skua, Auckland Island shags and sooty shearwater.

Activities

Snorkelling
Because access to the islands is only available through organised charter trips, a dive master will usually decide on the viability of snorkelling and the safest place to snorkel. Seals and sea lions can appear at any time and visibility is normally above 20 m.

Scuba-diving
Some of the charter boat trips to the Auckland Islands arrange some scuba-diving if weather conditions allow. Normally an onboard dive master will choose the most appropriate dive site on the day. Expect cold water and often rough conditions.

Walking
Tourism is controlled and only 600 visitors are allowed to the islands each year. There are only a few marked tracks and the undergrowth is hard going.

GLOSSARY

ascidian colonial sea squirt

benthic living on or in the seabed

biodiversity variety or numbers of plant and animal species in an area or habitat

biomass the combined weight of individuals or organisms of a defined species

brachiopod an ancient two-shelled organism not related to bivalve molluscs; sometimes known as lamp shells or ladies' toenail shells

bryozoan colonial invertebrate resembling clumps of moss, corals or seaweeds

chiton a limpet-like shell made up of eight sections and surrounded by a girdle

Continental shelf the shallow, flat underwater plain that extends from the shoreline around New Zealand, to around 130 m from where it drops steeply to the deep ocean

coralline algae stony or coral-like, usually pink; often covers rocks like paint

crustacean invertebrate animal with external skeleton; examples include shrimps, barnacles, crabs and crayfish

EEZ Exclusive Economic Zone – 200 nautical miles offshore all around New Zealand

endemic unique to a specified region or country

gastropod usually coiled seashell; examples include snails, limpets and paua

gorgonian type of colonial coral with eight-tentacled polyps

greywacke sedimentary sandstone formed between 270 and 65 million years ago

holdfast the part of kelp or seaweed attached to the reef

hydroid a colonial invertebrate usually with tree-like structure

intertidal zone the zone between high and low tides

invertebrate animal lacking backbone; examples include sponges, snails, crabs and sea stars

La Niña/El Nino weather phenomenon that occurs every 3–7 years. La Niña usually brings more easterly winds to northern New Zealand and El Nino a westerly pattern

larva (pl larvae) free-swimming stage in the life cycle of most marine life, after hatching from eggs

mollusc marine or land invertebrate with a soft body and often a hard shell; examples include snails, slugs, octopus, squid and bivalves

nudibranch a slug-like mollusc with external gills and no shell

plankton small plants and animals that drift on the ocean currents

polyp individual member of a colonial animal such as a coral; often equipped with stinging cells

sessile animals that fasten permanently to the seabed, such as barnacles and corals

triplefin small fish often found in rock pools

LIST OF SPECIES

Common names of plants and animals mentioned in the book are listed below under general groupings. Maori names, where known, are added in brackets, followed by the scientific name. It should be noted that the list by no means contains all of the species to be found in New Zealand marine reserves.

Seaweeds (Rimurimu)
Bloodweed, *Apophlaea sinclairii*
Bull kelp (Rimurapa, Rimuroa), *Durvillea antarctica*
Codium weed, *Codium fragile*
Common kelp, *Ecklonia radiata*
Coralline algae, *Corallina officinalis*
Eel grass, *Zosteria nana*, *Z. muelleri*
Flapjack, *Carpophylum flexuosum*, *C. maschalocarpum*
Giant bladder kelp, *Macrocystis pyrifera*
Grapeweed, *Caulerpa geminata*
Neptune's necklace, *Hormosira banksii*
Rimu weed, *Caulerpa brownii*
Sea lettuce, *Ulva lactuca*
Strap kelp, *Lessonia variegata*, *L. brevifolia*

Plants
Mangrove tree (Manawa), *Avicennia marina*

Sponges (Kopuputai)
Finger sponge (Kopuputai roa), *Callyspongia sp.*, *Raspailia sp.*
Encrusting sponges, *Aplysilla sulphurea*, *Aaptos alata*, *Crella incrustans*, *Polymastia fusca*
Yellow encrusting sponge, *Cliona celata*
Flask sponge, *Leucettusa sp.*
Golf ball sponge (Porotaka moana), *Tethya sp.*
Grey pillow sponge, *Ancorina alata*
Grey cup sponge, *Axinella tricalyciformis*

Coelenterates (sea pens, jellyfish, anemones, corals)
Bamboo coral, *Acanella sp.*
Black coral (Totara), *Intregra fiordensis*
Brown sea anemone, *Anthopleura aureoradiata*
Common sea anemone, *Anthothoe albocincta*
Dead man's fingers (Kotatea), *Alcyonium aurantiacum*
Hydroid, *Solanderia ericopsis*
Jewel anemone, *Corynactis haddoni*
Orange-striped anemone, *Cricophorus nutrix*
Plate coral, *Turbinaria sp*
Red hydrocoral (Punga kura), *Errina novaezelandiae*
Red Waratah anemone (Kotare moana), *Actinia tenebrosa*
Sea pen (Mui moana), *Sarcophyllum bollonsi*
Solitary cup coral (Pungatea), *Monomyces rubrum*
Starburst sea pen (Mui moana), *Kophobolemnon sp.*
Tube anemone, *Cerianthus bollonsi*
Wandering anemone (Humenga), *Phlyctenactis tuberculosa*
Zoanthid, *Parazoanthus sp.*

Bryozoans (lace corals, moss-like colonial invertebrates)
Blue-green bryozoan, *Bugula dentata*
Lace bryozoan (Punga tatari), *Hippellozoon novaezelandiae*

Molluscs (shells, octopus, squid, nudibranchs)

Apricot nudibranch (Rori), *Tritonia incerta*
Arabic volute (Pupu rore), *Alcithoe arabica*
Arrow squid (Ngu), *Nototodarus sloanii*
Black nerita snail (Peke, Matangarahu), *Nerita atramentosa*
Black-foot paua (Paua), *Haliotis iris*
Blue mussel, *Mytilus edulus aoteanus*
Cat's eye (Pupu, Ataata), *Turbo smaragdus*
Circular saw shell (Repo matamata), *Astraea heliotropium*
Clown nudibranch (Rori), *Ceratosoma amoena*
Coarse dosinia (Harihari), *Dosinia anus*
Cockle (Tuangi), *Austrovenus stutchburyi*
Cook's turban shell (Karaka), *Cookia sulcata*
Dark rock shell (Maihi, Ngaea), *Haustrum haustorium*
Dog cockle (Kuhakuha), *Tucetona laticostata*
Fan scallop (Kopakopa, tipa), *Chlamys zealandiae*
Gem nudibranch (Rori), *Dendrodoris denisoni*
Golden limpet (Ngakihi), *Cellana flava*
Gold-lined nudibranch (Rori), *Chromodoris aureomarginata*
Green-lipped mussel (Kuku), *Perna canaliculus*
Grooved fan shell, *Pallium convexum*
Harbour trough shell, *Mactra ovata*
Horn shell (Huamutu), *Zeacumanthus subcarinatus*
Horse mussel (Hururoa), *Atrina zelandica*
Jason nudibranch (Rori), *Jason mirabilis*
Kermadec spiny oyster, *Spondylus raoulensis*
Kermadec limpet, *Patella kermadecensis*
Kermadec royal top shell, *Tectus royanus*
Lemon nudibranch (Rori), *Dendrodoris citrina*
Lined whelk (Huamutu), *Buccinulum pallidum powelli*
Little black mussel, *Xenostrobus pulex*
Morning star shell (Tawera), *Tawera spissa*
Mud snail (Titiko), *Amphibola crenata*
Mud whelk, *Cominella glandiformis*
Mudflat top shell (Whetiko), *Diloma subrostrata*
Mudflat horn shell (Koeti), *Zeacumantus lutulentus*
Northern siphon whelk (Kakara nui), *Penion cuverianus*
Nut shell, *Nucula hartvigiana*
Opal top shell (Matangongo), *Cantharidus opalus*
Ostrich foot (Totorere, Kaikaikaroro), *Struthiolaria papulosa*
Oyster borer (Kaikai tio), *Lepsiella scobina*
Pacific oyster (Tio), *Crassostrea gigas*
Paper nautilus (Pupu tarakihi), *Argonauta nodosa*
Periwinkle (Ngaeti), *Nodilittorina antipodum, N.cincta*
Pink opal top shell, *Cantharidus purpureus*
Pipi (Pipi), *Paphies australis*
Queen scallop (Tipa), *Pecten novaezelandiae*
Ribbed mussel, *Aulocomya ater, A. maoriana*
Rock pool octopus (Wheke), *Octopus gibbsi*
Rock borer (Patiotio), *Barnea similis*
Rock oyster (Tio), *Crassostrea glomerata*
Royal helmet shell, *Semicassis royana*
Sand octopus (Wheke), *Octopus codiformis*
Shield shell (Rori), *Scutus antipodes*
Silky dosinia, *Dosinia lambata*
Snakeskin chiton (Papatua), *Chiton pelliserpentis*
Southern fan shell, *Chlamys dieffenbachi*
Southern file shell, *Lima zealandica*
Southern opal top shell, *Cantharidis capillaceus capillaceus*
Southern tiger shell, *Calliostoma granti*

Speckled whelk (Kawari), *Cominella adspersa*

Spindle cowrie, *Phenacovolva wakagamaensis*
Spotted top shell (Maihi), *Melagraphia aethiops*
Spotted whelk, *Cominella maculosa*
Tiger shell (Maurea), *Calliostoma tigris*
Toredo borer (Korotupa), *Teredo antarctica*
Trumpet shell (Awanui), *Charonia lampas*
Tuatua (Tuatua), *Paphies subtriangulata*
Turret shell (Papatai), *Maoricolpus roseus*
Variable nudibranch (Rori), *Aphelodoris luctuosa*
Verco's nudibranch (Rori), *Tambja verconis*
Virgin paua (Marapeka), *Haliotis virginea huttoni*
Warty sea hare (Rori), *Dolibrifera brazieri*
Wedge shell (Hanikura), *Macomona liliana*
Wellington nudibranch (Rori), *Archidoris wellingtonensis*
White rock shell (Hopetea), *Dicathais orbita*
Window oyster (Poro), *Anomia trigonopsis*
Yellow-foot paua (Hihiwa), *Haliotis australis*

Brachiopods
White brachiopod (Papa kura), *Liothyrella neozelandea*
Red brachiopod (Papa kura), *Magasella sanguinea, Calloria inconspicua*

Marine worms
Parchment worm, *Chaetopteris sp.*
Grey flatworm (Toke papa), *Thysanozoon brocchii*

Crustaceans (crabs, crayfish, shrimps, barnacles)
Acorn barnacle (Tiotio), *Balanomorpha sp.*
Banded coral shrimp, *Stenopus hispidis*
Black finger crab, *Ozius truncatus*
Common shrimp (Tarawera), *Palaemon affinis*
Common mud crab (Kairau), *Helice crassa*
Decorator crab (Papaka huna), *Notomithrax sp.*
Giant spider crab, *Jacquinotia edwardsii*
Green harbour crab, *Hemigrapsis crenulatus*
Half crab (Kawekawe), *Petrolisthes elongatus*
Mantis shrimp (Mana), *Lysiosquilla spinosa*
Packhorse or green crayfish (Koura uriuri), *Jasus verreauxi*
Paddle crab (Papaka), *Ovalipes catharus*
Purple rock crab (Papaka nui), *Leptograpsus variegatus*
Red rock crab (Papaka ura), *Plagusia chabrus*
Red crayfish (Koura moana), *Jasus edwardsii*
Rock pool hermit crab (Papaka moke), *Pagurus novaezelandiae*
Scampi, *Metanephrops challengeri*
Sea spider, *Pycnogonid sp.*
Slipper lobster, *Arctites antipodum*
Snapping shrimp (Korowhitiwhiti moana), *Alpheus novaezealandiae*
Stalk-eyed mud crab, *Macrophthalmus hirtipes*

Echinoderms (sea stars, sea urchins, sea cucumbers)
Biscuit sea star (Kapu ringa), *Pentagonaster pulchellus*
Brittle star (Weki huna), *Amphiura rosea*
Brown sea urchin, *Heliocidaris tuberculata*
Common sea urchin (Kina), *Evechinus chloroticus*
Crown of thorns sea star, *Acanthaster planci*
Cushion star (Kapu parahua), *Patriella regularis*
Firebrick sea star (Kapu ura), *Asterodiscus truncatus*
Heart urchin (Kina pakira), *Echinocardium cordatum*
Inflated cushion star (Kapu parahua rahi), *Stegnaster inflatus*

Palmer's diadema urchin, *Diadema palmeri*
Pencil urchin, *Phyllacanthus imperialis*
Pink urchin, *Tripneustes gratilla*
Red sea cucumber, *Ocnus sp.*
Rodolph's sea star, *Astrostole rodolphi*
Roger's sea urchin, *Centrostephanus rodgersii*
Sand dollar urchin (Kina papa), *Fellaster zelandiae*
Sea cucumber (Rori), *Stichopus mollis*
Snake star, *Astrobrachion constrictum*
Spiny sea star (Papatangaroa), *Coscinasterias muricata*
Strawberry sea cucumber, *Ocnus sp.*
Violet-blue diadema urchin, *Diadema sp.*
White sea urchin, *Pseudechinus huttoni*
White sea cucumber, *Ocnus brevidentis*

Tunicates (sea squirts)
Orange ascidian (Kaeo kura), *Pseudodistoma novaezelandiae*
Sea tulip (Kaeo), *Boltenia pachydermatina*
Giant salp, *Pyrosoma sp.*

Marine mammals (Dolphins, whales, seals and sea lions)
Blue whale, *Balaenoptera musculus*
Bottlenose dolphin, *Tursiops truncatus*
Bryde's whale, *Balaenoptera edeni*
Common dolphin (Aihe), *Delphinus delphis*
Dusky dolphin, *Lagenorhynchus obscurus*
Elephant seal (Ihu koropuka), *Mirounga leonina*
False killer whale, *Pseudorca crassidens*
Fin whale, *Balaenoptera physalis*
Hector's dolphin (Tupoupou), *Cephalorynchus hectori hectori*
Humpback whale (Paikea), *Megaptera novaeangliae*
Leopard seal (Popoiangore), *Hydrurga leptonyx*
Maui dolphin (Maui), *Cephalorynchus hectori maui*
Minke whale, *Balaenoptera acutorostrata*
New Zealand fur seal (Kekeno), *Arctocephalus forsteri*
New Zealand or Hooker's sea lion (Whakahao), *Phocarctos hookeri*
Orca (killer whale) (Kakahi), *Orcinus orca*
Pilot whale (long-finned), *Globicephala melaena*
Pilot whale (short-finned), *Globicephala macrorhynchus*
Sei whale, *Balaenoptera borealis*
Southern right whale, *Balaena glacialis*
Sperm whale (Paraoa), *Physeter macrocephalus*

Fish
Antarctic cod, *Notothenia microlepidota*
Banded wrasse (Tangahangaha), *Notolabrus fucicola*
Bigeye, *Pempheris adspersus*
Black angelfish, *Parma alboscapularis*
Black cod, *Paranotothenia magellanica*
Black-spotted goatfish, *Parupeneus signatus*
Blue knifefish, *Labracoglosso nitida*
Blue maomao (Maomao), *Scorpis violaceus*
Blue moki (Moki), *Latridopsis ciliaris*
Blue cod (Rawaru), *Parapercis colias*
Bluefish (Korokoropounamu), *Girella cyanea*
Blue shark , *Prionace glauca*
Bronze whaler shark (Horopekapeka), *Carcharhinus brachyurus*
Butterfish (Marari), *Odax pullus*
Butterfly perch (Oia), *Caesioperca lepidoptera*
Caramel drummer, *Girella fimbriata*
Combfish, *Coris picta*
Common conger eel (Ngoiro), *Conger verreauxi*
Common triplefin, *Forsterygion lapillum*
Cook's scorpionfish, *Scorpaena cookii*
Copper moki, *Latridopsis forsteri*
Crested blenny, *Parablennius laticlavius*
Crested weedfish, *Ericentrus rubrus*
Crimson cleanerfish, *Suexichthys aylingi*
Dwarf scorpionfish (Matuawhapuku), *Scorpaena papillosus*
Eagle ray (Whai manu), *Myliobatis*

List of Species **163**

tenuicaudatus
Electric Ray (Whai repo), *Torpedo fairchildi*
Elegant wrasse, *Anampses elegans*
Foxfish (Kotakota), *Bodianus sp.*
Galapagos shark, *Carcharhinus galapagensis*
Giant boarfish, *Paristiopterus labiosus*
Girdled wrasse, *Notolabrus cinctus*
Goatfish or red mullet (Ahuruhuru), *Upenichthys lineatus*
Gold-ribbon grouper, *Aulacocephalus temmincki*
Great white shark (Mango tuatine), *Carcharodon carcharius*
Green wrasse, *Notolabrus inscriptus*
Grey knifefish, *Bathystethus cultratus*
Grey moray eel, *Gymnothorax nubilus*
Grey mullet, *Mugilcephalus*
Grey drummer, *Kyphosus bigibbus*
Gurnard (Kumukumu), *Chelidonichthys kumu*
Half-banded perch, *Hypoplectrodes sp.*
Hapuku or groper (Hapuku), *Polyprion oxygeneios*
Jack mackerel or yellowtail (Hauture), *Trachurus novaezelandiae*
John dory (Kuparu), *Zeus faber*
Kahawai (Kahawai), *Arripis trutta*
Kelpfish (Hiwihiwi), *Chironemus marmoratus*
Kermadec demoiselle, *Chrysiptera rapanui*
Kermadec angelfish, *Parma kermadecensis*
Kingfish (Haku), *Seriola lalandi*
Koheru (Koheru), *Decapterus koheru*
Lavender lizardfish, *Synodus similis*
Leatherjacket (Kokiri), *Parika scaber*
Lionfish, *Pterois volitans*
Long-finned boarfish, *Zanclistius elevatus*
Long-finned eel (Kaiwharuwharu), *Anguilla ddieffenbachii*
Long-snouted butterflyfish, *Forcipiger flavissimus*
Long-tailed stingray (Whai), *Dasyatis thetidis*
Lord Howe coralfish, *Amphichaetodon howensis*
Mado (Mado), *Atypichthys latus*

Magpie perch, *Gonistius Bizonarius*
Maori chief, *Notothenia angustata*
Marblefish (Kehe), *Aplodactylus arctidens*
Moorish idol, *Zanclus cornutus*
Mosaic moray eel, *Enchelycore ramosa*
Mottled moray, *Gymnothorax prionodon*
Native trout (Kokopu), *Glaxias fasciatus*
Northern conger eel (Ngoiro), *Conger wilsoni*
Northern kahawai, *Arripis xylabion*
Northern scorpionfish (Matuawhapuku), *Scorpaena cardinalis*
Northern splendid perch, *Callanthias australis*
Notch-head marblefish, *Aplodactylus etheridgii*
Oblique-swimming triplefin, *Obliquichthys maryannae*
Orange wrasse, *Pseudolabrus luculentus*
Pacific gregory, *Stegastes fasciolatus*
Painted moki, *Cheilodactylus ephippium*
Parore (Parore), *Girella tricuspidata*
Pink maomao (Matata), *Caprodon longimanus*
Pipefish, *stigmatophora macropterygia*
Piper (Ihe, Takeke), *Hyporhamphus ihi*
Porae (Porae), *Nemadactylus douglasii*
Red lizardfish, *Synodus doaki*
Red moki (Nanua), *Cheilodactylus spectabilis*
Red pigfish (Pakurakura), *Bodianus unimaculatus*
Red-banded perch, *Hypoplectrodes huntii*
Rock cod (Taumaka), *Lotella rhacinus*
Sand flounder (Patiki), *Rhombosolea plebeia*
Sandager's wrasse, *Coris sandageri*
Scarlet wrasse, *Pseudolabrus miles*
Sea horse (Manaia, Kiore moana), *Hippocampus abdominalis*
Sea perch (Pohuiakaroa), *Helicolenus percoides*
Seven-gilled shark (Tuatini), *Notorhynchus cepedianus*
Sharp-nosed puffer, *Canthigaster callisterna*
Short-tailed stingray (Whai, Oru), *Dasyatis brevicaudata*

Silver drummer, *Kyphosus sydneyanus*
Single-spot demoiselle, *Chromis hypsilepis*
Slender roughy (Puramorehu), *Optivus elongatus*
Snapper (Tamure), *Pagrus auratus*
Southern pigfish (Purumorua), *Congiopodus leucopaecilus*
Southern splendid perch, *Callanthias allporti*
Speckled moray eel, *Gymnothorax obesus*
Spiny sea dragon, *Solegnathus spinosissimus*
Splendid hawkfish, *Cirrhitus splendens*
Spotted black grouper, *Epinephelus daemelii*
Spotted dog shark (Pekepeke), *Squalus acanthias*
Spotty (Paketi/Pakirikiri), *Notolabrus celidotus*
Stargazer (Kutoro), *Geniagnus monoterygius*
Sunfish (Ratahuihui), *Mola mola*
Sweep (Hui), *Scorpis lineolatus*
Tarakihi (Tarakihi), *Nemadactylus macropterus*
Tattooed rockskipper, *Entomacrodus niuafoouensis*
Telescopefish (Koihi), *Mendosoma lineatum*
Toadstool grouper, *Trachypoma macracanthus*
Trevally (Araara), *Pseudocaranx dentex*
Trumpeter (Kohikohi), *Latris lineata*
Trumpetfish, *Aulostomus chinensis*
Two-spot demoiselle, *Chromis dispilus*
Whitebait (Kokopu, Inunga), *Galaxias sp.*
Yellow demoiselle, *Chromis fumea*
Yellow moray eel (Puharakeke), *Gymnothorax prasinus*
Yellow-back triplefin, *Forsterygion flavonigrum*
Yellow-banded perch, *Acanthistius cinctus*
Yellow-eyed mullet (Aua), *Aldrichetta forsteri*
Yellow-striped boarfish, *Evistias acutirostris*
Wavy-lined perch, *Lepidoperca tasmanica*

Birds
Antarctic tern (Tara), *Sterna vittata*
Antarctic prion (Whiroia), *Pachyptila desolata*
Auckland Island shag (Koau), *Phalacrocorax colensoi*
Australasian gannet (Takapu), *Morus serrator*
Banded dotterel (Tuturewhatu), *Charadrius bicinctus bicinctus*
Banded rail, *Gallirallus philippensis*
Bellbird (Korimako), *Anthornis melanura*
Bittern (Matuku), *Botaurus poiciloptilus*
Black shag (Kawau), *Phalacrocorax carbo*
Black-backed gull (Karoro), *Larus dominicanus*
Black-capped petrel, *Pterodroma cervicalis*
Buller's shearwater, *Puffinus bulleri*
Cape pigeon (Titore), *Daption capense*
Caspian tern (Tara-nui), *Hydroprogne caspia*
Diving petrel (Kuaka), *Pelecanoides urinatrix*
Eastern bar-tailed godwit (Kuaka), *Limosa lapponica*
Erect-crested penguin, *Eudyptes sclateri*
Fairy prion (Titi-wainui), *Pachyptila turtur*
Fairy tern (Tara-iti), *Sterna nereis*
Fantail (Piwakawaka), *Rhipidura fuliginosa*
Fernbird (Matata), *Bowdleria punctata*
Fiordland crested penguin (Tawaki), *Eudyptes pachyrhynchus*
Flesh-footed shearwater, *Puffinus carneipes*
Fluttering shearwater (Pakaha), *Puffinus gavia*
Fulmar prion, *Pachyptila crassirostris*
Gibson's albatross, *Diomedea gibsoni*
Grey ternlet or grey noddy, *Procelsterna cerulea*
Grey-faced petrel (Oi), *Pterodroma macroptera*
Grey-headed mollymawk, *Thalassarche chrysostoma*
King shag (Kawau-pateketeke),

Phalacrocorax carunculatus
Kingfisher (Kotare), *Halcyon sancta*
Lesser knot (Huahou), *Calidris canutus*
Light-mantled sooty albatross, *Phoebetria palpebrata*
Little blue penguin (Korora), *Eudyptula minor*
Little spotted kiwi, *Apteryx oweni*
Little shag (Kawau-paka), *Phalacrocorax melanoleucos*
Longtail Cuckoo (Koekoea), *Eudynamys taitensis*
Mallard duck, *Anas platyrhynchos*
Masked (blue-faced) booby, *Sula dactylatra*
Northern giant petrel (Panguruguru), *Macronectes halli*
New Zealand dotterel, *Charadrius obscurus*
New Zealand pipit (Pihoihoi), *Anthus novaeseelandiae*
Pied shag (Kuruhiruhi), *Phalacrocorax varius*
Pied stilt (Poaka), *Himantopus leucocephalus*
Pukeko (Pukeko), *Porphyrio melanotus*
Pycroft's petrel, *Pterodroma pycrofti*
Red-billed gull (Tarapunga), *Larus novaehollandiae*
Red-tailed tropic bird (Amokura), *Phaethon rubricauda*
Reef heron (Matuku moana), *Egretta sacra*
Rock pigeon, *Columba livia*
Rockhopper penguin, *Eudyptes chrysocome*
Royal albatross (Toroa), *Diomedea epomophora*
Sandpiper, *Calidris ferruginea*
Shining cuckoo (Pipiwharauroa), *Chrysecoccyx lucidus*
Sooty shearwater (Titi), *Puffinus griseus*
South Island oystercatcher (Toroa), *Haematopus finschi*
Southern skua (Hakoakoa), *Catharacta lonnbergi*
Spotless crake (Putoto), *Porzana tabuensis*
Spotted shag (Parekareka), *Phalacrocorax punctatus*
Tui, *Brosthemadera novaesseelandiae*
Variable oystercatcher (Torea), *Haematopus unicolor*
Wandering albatross (Toroa), *Diomedea exulans*
Wax-eye (Pihipihi), *Zosterops lateralis*
Weka, *Gallirallus australis*
White-capped mollymawk, *Thalassarche steadi*
Welcome swallow, *Hirundo neoxena*
White-capped noddy, *Anous tenuirostris*
White-chinned petrel, *Procellaria aequinoctialis*
White-faced heron, *Ardea novaehollandiae*
White-faced storm petrel (Takahi-kare-moana), *Pelagodroma marina*
White-flippered penguin, *Eudyptula albosignata*
White-fronted tern (Tara), *Sterna striata*
White-headed petrel, *Pterodroma lessoni*
Wrybill, *Anarhynchus frontalis*
Yellow-eyed penguin (Hoiho), *Megadyptes antipodes*
Yellowhammer, *Emberiza citrinella*

REFERENCES

Kermadec Islands Marine Reserve

Department of Conservation(DoC) (1991). *The Kermadec Islands*. DoC.

Frances, M P, Grace, R V, & Paulin, C D (1987). Coastal fishes of the Kermadec Islands. *NZ Journal of Marine and Freshwater Research, 21*:1–13.

Francis, M, & Nelson, W (2003). Chapter 25, Kermadec Islands. *The Living Reef*. Nelson: Craig Potton Publishing.

Gabites, B (1993). Island of dreams. *New Zealand Geographic Magazine, 19*: 76–98.

Schiel, D R, Kingsford, M J, & Choat, J H (1986). Depth distribution and abundance of benthic organisms and fishes at the subtropical Kermadec Islands. *NZ Journal of Marine and Freshwater Research, 20*: 521–535.

Sykes, W R (1974). The Kermadecs. *New Zealand's Nature Heritage, 4*: 1641–1647.

Mimiwhangata Marine Park

Ballantine W J, Grace R V, & Doak W T (1973). *Mimiwhangata marine report*. City: Turbott & Halstead for NZ Breweries Ltd.

Denny C M, & Babcock R C (2002). *Fish survey of the Mimiwhangata Marine Park*. DoC.

Grace R V (1978). *Monitoring of the marine environment*. NZ Breweries Trust.

Grace R V, & Kerr V C (2002). *Mimiwhangata Marine Park draft report*. DoC.

Grace R V & Kerr V C (2003). *Mimiwhangata marine monitoring programme*. DoC.

Kerr, V C & Grace, R V (2005). *Intertidal and subtidal habitats of Mimiwhangata Marine Park and adjacent shelf*. DoC Research and Development Series 201. DoC.

Poor Knights Islands Marine Reserve

Ayling, T & Schiel, D (2003). Chapter 27, Poor Knights Islands. In Francis, M, & Nelson, W (2003). *The Living Reef*. Nelson: Craig Potton Publishing.

Denny, C M, Willis, T J & Babcock, R C (2003). *Effects of Poor Knights Islands Marine Reserve on demersal fish populations*. DoC Science Internal Series 142.

DoC (1993). Poor Knights Nature Reserve and Marine Reserve brochure. DoC.

Cape Rodney–Okakari Point Marine Reserve

Ayling, A M, Cumming, A, & Ballantine, W J (1981). Map of shore and subtidal habitats of the Cape Rodney to Okakari Point Marine Reserve, North Island, New Zealand in 3 sheets. Wellington: Dept of Lands and Survey.

Cole, R G, & Keuskamp, D (1998). Indirect effects of protection from exploitation: patterns from populations of *Evechinus chloroticus* (Echinoidea) in northeastern New Zealand. *Marine Ecology Progress Series, 173*:215–226.

Kelly, S, Scott, D, & MacDiarmid, A B (2002). The value of a spillover fishery for spiny lobsters around a marine reserve in northern New Zealand. *Coastal Management, 30*:153–166.

Kelly, S, Scott, D, MacDiarmid, A B, & Babcock, R C (2000). Spiny lobster, *Jasus edwardsii*, recovery in New Zealand marine reserves. *Biological Conservation, 92*:359–369.

Willis, T J, Millar, R B, & Babcock, R C (2003). Protection of exploited fish in temperate regions: High density and biomass of snapper *Pagrus auratus* (Sparidae) in northern New Zealand marine reserves. *Journal of Applied Ecology,* 40:213–227.

Tawharanui Marine Park
Grace, R V (1978). *Tawharanui marine monitoring programme.* Report prepared for the Auckland Regional Council (ARC).

Grace, R V (1981). *Tawharanui marine monitoring programme.* Report prepared for the ARC.

Grace, R V (1991). Tawharanui marine monitoring programme. Report prepared for the ARC.

Willis, T J, Millar, R B, & Babcock, R C (2003). ibid.

Long Bay–Okura Marine Reserve
ARC. Long Bay Regional Park (brochure). ARC.

Babcock, Creese and Walker (2000). *The Long Bay Monitoring Programme.* Sampling Report.

DoC (1995). Long Bay–Okura Marine Reserve brochure. DoC.

Motu Manawa (Pollen Island) Marine Reserve
DoC (1997). Motu Manawa (Pollen Island) Marine Reserve, Waitemata Harbour, Auckland brochure. DoC.

Sivaguru, K, & Grace, R (2002). Benthos and sediments of Motu Manawa (Pollen Island) Marine Reserve. *Auckland Conservancy Technical Series*, DoC.

Te Matuku Marine Reserve
Hayward, B W, Stephenson, A B, Morley, M S, Smith, N, Thompson, F, Blom, W, Stace, G, Riley, J L, Prasad, R, & Reid, C (1997). Intertidal Biota of Te Matuku Bay, Waiheke Island, Auckland. *Tane*, 36:67–84.

DoC (2005). Te Matuku Marine Reserve, Waiheke Island brochure. DoC.

Te Whanganui-a-Hei Marine Reserve
Haggitt, T & Kenny, S (2004). *Te Whanganui-a-Hei Marine Reserve Biological Monitoring Plan.* Coastal and Aquatic Systems Ltd.

Kelly, S, Scott, D, MacDiarmid, A, & Babcock, R (1999). Spiny lobster, *Jasus edwardsii*, recovery in New Zealand marine reserves, *Biological Conservation*, 92 (2000):359–369.

Willis, T J, Millar, R B, & Babcock, R C (2003). ibid.

DoC (1997). Te Whanganui-a-Hei Cathedral Cove Marine Reserve brochure. DoC.

Mayor Island (Tuhua) Marine Reserve
Prebble, G K (1971). 'Tuhua' Mayor Island. Ashford-Kent & Co.

Young, K, Ferreira, S, Jones, A, & Gregor, K (2004). *Reef fish response to full and partial protection in Tuhua Marine Reserve.* Tauranga: Marine Studies Department, Bay of Plenty Polytechnic.

Sugar Loaf Islands Marine Protected Area
DoC (1999). Nga Motu Sugar Loaf Islands Marine Protected Area brochure. DoC.

Te Tapuwae o Rongokako Marine Reserve
Freeman, D (2005). Personal communication on progress report on PhD study.

Te Angiangi Marine Reserve
DoC. Te Angiangi Marine Reserve Operational Plan. April 2003. DoC.

DoC (1997). Te Angiangi Marine Reserve brochure. DoC.

Kapiti Marine Reserve
DoC (1998). Kapiti Island Conservation Management Plan. DoC.

NIWA Survey of Kaimoana, 1999–2000. NIWA.

Blue Cod Study, 2003–2004. Victoria University.

Study of Kapiti Marine Reserve, 1998–2000. Victoria University.

Westhaven (Te Tai Tapu) Marine Reserve

DoC (1994). Westhaven Inlet Marine Reserve brochure. DoC.

Davidson, R J (1990). *A Report on the Ecology of Whanganui Inlet*. North-West Nelson.

Tonga Island Marine Reserve

Davidson, R (2001). *Tonga Island Marine Reserve: Proposed protocol for ongoing subtidal biological monitoring*. Research, Survey and Monitoring Report No. 316. DoC.

DoC (1999). Tonga Island Marine Reserve brochure. DoC.

Horoirangi Marine Reserve

Bradstock, M & Gordon, D (1983). Coral-like bryozoan growths in Tasman Bay and their protection to conserve fish stocks. *NZ Journal of Marine and Freshwater Research* 17:159–63.

Cole, N, Grange, K, Morrisey, D, & Glenduan (2003). *Survey, marine habitats, fish and benthic species within a proposed marine reserve, North Nelson*. NIWA.

Johnson, D, & Haworth, J (2004). *Hooked: The story of the New Zealand fishing industry*. Christchurch: Hazard Press.

Long Island–Kokomohua Marine Reserve

Davidson, R J (2004). Long Island–Kokomohua Marine Reserve, Queen Charlotte Sound: 1992–2003. *Survey and Monitoring Report No. 343*. Nelson: DoC.

DoC (1993). Long Island–Kokomohua Marine Reserve, Long Island Scenic Reserve brochure. DoC.

Pohatu Marine Reserve

DoC (2000). Banks Peninsula Marine Areas brochure. DoC.

Davidson, R, Barrier, R, & Pande, A (2001). Pohatu Marine Reserve Baseline Survey. DoC.

Davidson, R & Abel, W (2002). Second sampling of Pohatu Marine Reserve, Flea Bay, Banks Peninsula (interim report). DoC.

Fiordland Marine Reserves

Guardians of Fiordland's Fisheries and Marine Environment (2003). *Beneath the reflections: Fiordland marine conservation strategy*. GOFF.

Wing, S (2003). Chapter 30, Fiordland. *The Living Reef*. Nelson: Craig Potton Publishing.

Te Wharawhara (Ulva Island) Marine Reserve

Peat, N. (2000). *Stewart Island: A Rakiura Ramble*. Dunedin: University of Otago Press.

DoC (2004). Ulva Island/Te Wharawhara Marine Reserve brochure. DoC.

Auckland Islands (Motu Maha) Marine Reserve

Schiel, D & Kingsford, M (2003). Chapter 31, Subantactic Islands. *The Living Reef*. Nelson: Craig Potton Publishing.

FURTHER READING

Books and Periodicals:

Adams, N (1997). *Common seaweeds of New Zealand*. Christchurch: Canterbury University Press.

Powell, A W B (1998). *Powell's native animals of New Zealand*. Auckland: David Bateman.

Baker, A (1990). *Whales and dolphins of New Zealand and Australia: an identification guide*. Wellington: Victoria University Press.

Ballantine, W (1991). *Marine reserves for New Zealand*. University of Auckland, Leigh Marine Laboratory Bulletin No. 25.

Batson, P (2003). *Deep New Zealand*. Christchurch: Canterbury University Press.

Clover, C (2004). *The end of the line*. London: Ebury Press.

Dawson, S, & Slooten, E (1996). *Down-under dolphins: The story of Hector's dolphin*. Christchurch: Canterbury University Press.

Doak, W (1971). *Beneath New Zealand seas*. Wellington: Reed.

Doak, W (1972). *Fishes of the New Zealand region*. Auckland: Hodder and Stoughton.

Edney, G (2001). Poor Knights wonderland. *Sea Tech Ltd/Dive New Zealand Magazine*. Auckland.

Enderby T & J (1998). *Goat Island Marine Reserve*. Self-published.

Enderby, T & J (2002). *Diving and snorkeling New Zealand*. Melbourne: Lonely Planet Publications.

Francis, M (1996). *Coastal fishes of New Zealand: an identification guide*. Auckland: Reed.

Francis M, & Andrew, N (2003). *The living reef: The ecology of New Zealand's rocky reefs*. Nelson: Craig Potton Publishing.

Guardians of Fiordland's Fisheries and Marine Environment (2003). *Beneath the reflections: Fiordland marine conservation strategy*. Invercargill: GOFF.

Hall-Jones, J (1990). *Fiordland explored*. Invercargill: Craig Printing Co.

Medway, D (2002). *Sea and shore birds of New Zealand*. Auckland: Reed.

Morton, J (2004). *Seashore ecology of New Zealand and the Pacific*. Auckland: David Bateman.

Morton, J, & Miller, M (1978). *The New Zealand seashore*. Auckland: Collins.

Parkinson, B (2000). *Field guide to New Zealand seabirds*. Auckland: New Holland (NZ).

Paulin, C, & Roberts, C (1992). *The rockpool fishes of New Zealand*. Auckland: Museum of New Zealand.

Peat, N (2000). *Stewart Island: A Rakiura ramble*. Dunedin: University of Otago Press.

Reed, A H (republished 2004). *The Four Corners of New Zealand*. Auckland: New Holland (NZ).

Ryan, P, & Paulin, C (1998). *Fiordland underwater*. Auckland: Exisle.

Soper, M F (1984). *Birds of New Zealand and outlying islands*. Christchurch: Whitcoulls Publishers.

Spencer, H, & Willan, R (1995). *The marine fauna of New Zealand: Index to the Fauna 3, Mollusca*. New Zealand Oceanographic Institute Memoir 105.

Stace, G (1997). *What's on the beach: A guide to coastal marine life*. Auckland: Penguin Books (NZ).

Stace, G (1998). *What's around the rocks: A simple guide to the rocky shore*. Auckland: Penguin Books (NZ).

Websites:

Auckland Regional Council (ARC), www.arc.govt.nz

Convention on International Trade in Endangered Species of Wild Fauna and Flora (CITES), www.cites.org

Department of Conservation, www.doc.govt.nz

Jenny & Tony Enderby, www.enderby.co.nz

Experiencing Marine Reserves, www.emr.org.nz

Guardians of Fiordland's Fisheries & Marine Environment Inc. www.fiordland-guardians.org.nz

Leigh Marine Laboratory, www2.auckland.ac.nz/Leigh

Ministry for the Environment, www.mfe.govt.nz

National Institute of Water and Atmospheric Research (NIWA), www.niwa.co.nz

No-take Marine Reserves www.marine-reserves.org.nz

Royal Forest and Bird Protection Society of New Zealand Inc. www.forestandbird.org.nz

INDEX

Acheron Passage 144
albatross 157, 158
anemone
  brown sea 57
  common sea 27, 44, 51, 56, 72, 83, 109–110, 121
  jewel 27, 44, 51, 72, 83, 121
  orange-striped 131, 150
  red Waratah 33
  wandering 116, 150
angelfish
  black 27, 33, 34, 77
  Kermadec 22
Arabic volute 27, 67, 110, 150
Aramoana 92, 95
ascidian 27, 28, 34, 41, 44, 51, 56, 73, 77, 84, 89, 94, 121, 126, 133
Australasian gannet 28, 36, 45, 73, 79, 90, 94, 100, 106, 111, 122

Banks Peninsula 123–7
barnacle, acorn 66
bellbird 52, 111, 152
bigeye 44
bittern 65, 68, 106
Blackhead 92, 95
black nerita snail 21, 26, 56, 72, 109
Bligh Sound 138
bloodweed 50
bluefish 21
boarfish
  giant 28, 44, 51, 79
  long-finned 36, 79
  yellow-striped 22
brachiopod 116, 133, 142, 146, 149–50
Bradshaw Sound 140
brittle star 41, 78, 116
bryozoan 34–5, 41, 51, 56, 79, 84, 94, 116, 133, 144
butterfish 34, 42, 89, 93–4, 99, 116, 120, 126, 131, 150

cat's eye 26, 41, 50, 56–7, 66–7, 78, 88, 93, 105, 109, 115, 121, 125
Charles Sound 139

circular saw shell 133
cockle 57, 61–2, 66–7, 105, 111, 140, 144
cod
  Antarctic 157
  black 157
  blue 42, 78, 84, 89, 99, 110, 116, 118, 120–1, 126, 133, 144–5, 150
  red 133
codium weed 56, 132
combfish 34
Cook, James 31, 76, 119, 129, 144–5, 148
Cook's turban shell 27, 43, 50–1, 56, 72, 93, 99, 110, 115–6, 121, 133, 150
coral
  bamboo 28
  black 15, 23, 28, 79, 128, 133–4, 136, 138–40, 142, 144, 146
  plate 23
  solitary cup 35, 79, 99, 133
coralline algae 22, 41, 56, 83, 93, 109, 116, 126, 132, 142
crab
  black finger 42
  decorator 41, 50, 137
  giant spider 157
  green harbour 105
  half 41, 50, 56
  hermit 41, 50, 56–7, 88, 116, 121, 133
  mud 57, 62, 66–7, 104–5
  paddle 26, 57, 105, 110, 116
  purple rock 150
  red rock 42, 50, 57, 78, 105
crayfish
  red 11, 27, 35, 39–0, 42–4, 47, 49–51, 56, 68, 71–2, 77–8, 84–5, 88–90, 93–4, 99–100, 109–10, 118, 120–1, 126, 128, 133, 139–40, 147, 150
  packhorse or green 27–8, 35, 44, 50, 78
crested weedfish 27, 93
crimson cleanerfish 34
cuckoo
  long-tailed 28
  shining 28
cushion star 41, 50, 67, 72, 105, 109–10, 116, 131

dead man's fingers 51, 73
demoiselle 21–2, 34, 36, 44, 73, 78–9, 84
dog cockle 26, 116
dolphin
   bottlenose 23, 28, 36, 44, 52, 73, 79, 84, 89, 94, 100–1, 117, 122, 134, 150
   common 28, 36, 44, 73, 79, 84, 94, 122
   dusky 94, 111, 122
   Hector's 15, 111, 122–3, 126–7
   maui 15, 84
dosinia 57, 67, 116
dotterel
   banded 106
   New Zealand 28, 52, 63
Doubtful Sound 141, 143
drummer
   caramel 21
   grey 21
   silver 11, 43–4, 50, 84, 99
Dusky Sound 141, 145

eel
   conger 44, 133
   long-finned 104
   moray 7, 27, 35, 44, 73, 78
eel grass 67–8, 104–5
elephant seal 89, 157

fantail 45, 68
fernbird 63, 68, 106
flounder 57, 62, 67, 104–5, 137
foxfish 28

gannet, *see* Australasian gannet
giant salp 33
glassworts 61–68
goatfish (red mullet) 26, 28, 34, 43, 57, 72, 77, 84, 94, 99, 110, 116
   black-spotted 22
godwit 62, 68, 94, 106
grapeweed 50
grey ternlet or grey noddy 23
grouper (hapuku) 7, 28, 32, 94, 120, 126, 140, 144, 150
   gold-ribbon 22, 35
   spotted black 15, 18, 22, 28, 34, 100
   toadstool 22, 35
gull
   black-backed 28, 45, 85, 90, 100, 106, 158
   grey-faced 36, 52, 85
   red-billed 28, 36, 45, 85, 90, 100, 106, 157
gurnard 27–8, 51, 84, 89, 94, 99, 116

Hahei 69
hapuku *see* grouper
harbour trough shell 62, 67
hawkfish 22
hermit crab 41, 50, 56–7, 88, 116, 121, 133
heron
   reef 28, 45, 85, 94, 100
   white-faced 45, 57, 67, 68, 94, 106, 111
horn shell 62, 67, 93, 105
hydroid 27–8, 34–5, 44, 51, 73, 77, 84, 89, 94, 121, 142

jack mackerel (yellowtail) 27, 34, 42, 44, 72–3, 78–9, 84, 99, 116
John dory 51

kahawai 28, 44, 51, 67, 78, 81, 84, 99, 104–5
   northern 22
kelp
   bull 125, 131, 150, 156
   common 27, 42, 51, 56, 67, 77, 78, 83, 88, 94, 131–2
   giant bladder 126, 131, 145, 150, 156
   strap 34, 131
kelpfish 27, 42, 50, 72, 77, 83, 89, 99, 121
Kermadec demoiselle 22
Kermadec royal top shell 22
Kermadec spiny oyster 23
killer whale *see* orca
kingfish 22, 27, 32, 34, 42, 44, 72–3, 79, 81, 84, 100
kingfisher 28, 45, 57, 63, 67–8, 94, 106
kiwi, little spotted 100, 122
knifefish
   blue 21
   grey 21
koheru 27, 34, 44, 73, 79, 84

leatherjacket 27, 34, 44, 72, 77, 89, 99, 110, 116, 121, 126, 133, 150
leopard seal 89, 134, 157
lesser knot 62, 68

Index **173**

limpet
   golden 88, 93
   Kermadec 21
lionfish 22
little spotted kiwi 100, 122
lizardfish 34
long-snouted butterflyfish 22
Lord Howe coralfish 22, 35, 78

mado 22, 51
mallard duck 45
mangrove 56–7, 59, 61–2, 64, 66–8
maomao
   blue 22, 27, 33–4, 38, 42–4, 51, 73, 78, 84, 100, 150, 157
   pink 22, 28, 33–6, 78–9
marblefish 22, 27, 43, 72, 83, 89, 93, 99, 116, 131
   notch-head 21
marine weather 17
masked (blue-faced) booby 23
Milford Sound 128–9, 134–6
moki
   blue 78, 84, 89, 93–4, 120–1, 126
   copper 84
   painted 22
   red 27, 34, 42, 50, 72, 77, 89, 110, 116
Moorish idol 22
morning star shell 26, 116
mud snail 61–2, 66–7, 104–5, 111
mullet
   grey 56
   yellow-eyed 56–7, 62, 67, 105, 111
mussel
   blue 109, 125, 131–2, 149, 157
   green-lipped 57, 67, 72, 83, 116, 121, 125
   horse 26, 57, 67, 110, 116, 121, 132, 134
   little black 66–7, 72, 125
   ribbed 105, 131, 157

Nelson 113–16
Neptune's necklace 41, 56, 67, 88, 93, 105
New Zealand fur seal 36, 73–4, 81, 84–5, 89, 94, 100, 101, 107, 111, 116, 122–3, 126–7, 129, 131, 134, 149–51, 157
New Zealand (Hooker's) sea lion 89, 134, 150–1, 153, 155–7
New Zealand pipit 45

nudibranch
   apricot 51, 73
   clown 41, 51, 84, 94
   gem 51
   gold-lined 94, 132
   Jason 35, 51, 73, 84, 121, 132
   lemon 50
   variable 51
   Verco's 34
   Wellington 110, 116, 132
nut shell 62, 67

octopus 7, 42, 50, 79, 99–100, 126, 150
orca (killer whale) 23, 28, 36, 44, 52, 73, 79, 84, 89, 94, 100, 116, 122, 157
ostrich foot 67, 110
oyster 57, 61, 66–7, 72, 105, 115–6, 15
oysterborer 26, 41, 56, 67
oystercatcher 28, 45, 52, 57, 62, 68, 90, 94, 100, 106, 111

Pacific gregory 22
paper nautilus 79, 89
parchment worm 51
parore 38, 44, 46, 56–7, 62, 89
paua 72, 88, 99, 116, 120–1, 125, 147, 149, 150, 157
penguin
   erect-crested 157
   Fiordland crested 134
   little blue 28, 36, 45, 52, 73, 85, 94, 100–1, 111, 122, 134, 151
   rockhopper 157
   white-flippered 123, 125–6
   yellow-eyed 123, 126, 151, 157
perch
   butterfly 28, 35–6, 44, 73, 78, 89, 94, 100, 121, 133
   half-banded 35
   magpie 84
   red-banded 84, 94, 100, 126, 133
   sea 84, 89, 94, 100, 121, 133
   splendid 34, 79, 133, 146
   wavy-lined 133
   yellow-banded 22, 35
periwinkle 26, 41, 56, 67, 105, 125
petrel
   black-capped 23
   diving 36, 85, 100, 157
   northern giant 157
   Pycroft's 36

storm 36, 85, 157
   white-chinned 157
   white-headed 157
pied stilt 28, 57, 68, 94
pigeon
   Cape 158
   rock 58
pigfish
   red 22, 28, 36, 44, 73
   southern 126, 150
pipefish 150
piper 27, 42
pipi 57, 61, 62, 67, 105, 144
porae 33–5, 78
prion
   fairy 36, 126
   fulmar 158
   Antarctic 158
pukeko 62

rail, banded 63, 106
Raoul Island 19–22
ray
   eagle 27–8, 43, 51–2, 72, 99, 105, 116
   electric 51, 150
   see also stingray
red hydrocoral 128, 133–4, 138–40, 142–3, 146
red-tailed tropic bird 23
rock shell 41, 56, 105
rock borer 50
royal helmet shell 34

sandpiper 62, 68
scallop
   fan 131, 134, 150
   queen 26, 51, 57, 67, 110, 116, 121, 132, 134, 145–6, 150
scorpionfish 34
   Cook's 22
   dwarf 51, 73, 84, 89, 94
   northern 33
sea cucumber 78, 110, 116, 121, 131–2, 146
sea horse 8, 150
sea lettuce 27, 33, 56, 72, 77, 105, 131, 14
seamounts 14
sea pen 134, 142, 145–6
sea spider 131
sea squirt 116, 121

sea star
   biscuit 125, 150
   crown-of-thorns 22
   firebrick 34
   Rodolph's sea star 22
   spiny 41, 67, 116, 125, 131
sea tulip 121, 125–6, 150
shag
   Auckland Island 158
   black 28, 85, 100
   king 122
   little 100, 106
   pied 28, 36, 45, 79, 122, 151
   spotted 116, 126
shark
   blue 100, 133
   bronze whaler 32, 36, 100
   Galapagos 22
   great white 157
   seven-gilled 133, 144
   spotted dog 116, 133, 137
sharp-nosed puffer 22, 78
shearwater
   Buller's 36
   flesh-footed 36
   fluttering 36, 85, 100, 111
   sooty 36, 84, 100, 151, 158
shield shell 88, 110, 132
shrimp 41, 50, 56, 131
   banded coral 35
   mantis 105–6
slender roughy 44
slipper lobster 35, 44
snakeskin chiton 26
snake star 133, 142, 150
snapper 12, 26–7, 32, 34, 38–43, 49–51, 56–7, 62, 67, 71–2, 77, 84, 100, 104–5, 110, 116, 121
southern file shell 134
southern skua 158
spindle cowrie shell 36
spiny sea dragon 136
sponge
   finger 28, 44, 51, 56, 84, 116
   flask 35
   glass 143
   golfball 28, 36, 44, 51, 56, 84, 116
   grey cup 133
spotless crake 62, 106
spotty 27, 42–3, 50, 56–7, 67, 72, 77, 83,

89, 93, 99, 110, 116, 121, 126, 132
squid, arrow 157
stargazer 51, 94, 137
Stewart Island 147–9
stingray 27–8, 34, 36, 44, 51, 57, 62, 122
sunfish 34
sweep 27, 43, 51, 73, 78, 89

tarakihi 28, 33, 51, 84, 89, 94, 99–100, 110, 116, 120–22, 126, 140, 150
tattooed rockskipper 21
telescopefish 131, 157
tern
   Antarctic 158
   Caspian 28, 45, 58, 63, 100, 106
   fairy 23
   white-fronted 28, 36, 45, 52, 57, 63, 85, 100, 106, 122, 158
tiger shell 27, 51, 72, 77, 84, 99, 133
   southern 133, 157
top shell
   mudflat 62
   opal 27, 149–50, 157
   spotted 50, 93
toredo mollusc 132
trevally 22, 28, 32, 43, 51, 77, 84
triplefin
   common 131
   oblique-swimming 131
   yellow-back 132
trumpeter 126, 150, 157
trumpetfish 22
trumpet shell 27, 41, 51, 77, 133
tuatua 26, 57
tube worms 57, 67, 121, 131
tui 73
turret shell 67, 131
turtles 21, 34

urchin
   brown sea 21
   common sea 11, 22, 41–3, 50, 56, 71–2, 77–8, 83, 88, 93, 99, 105, 109, 116, 120–1, 125, 131
   heart 57, 116
   Palmer's diadema 23, 36
   pencil 21–2
   pink 21, 131
   Roger's sea 21, 34, 78
   sand dollar 57, 67, 116
   violet-blue diadema 23
   white sea 131–2

warty sea hare 21
waxeye 28
wedge shell 62, 67, 105
weed
   codium 56, 132
   flapjack 27, 42, 56, 67, 72, 77, 83, 88, 93, 99, 109–11, 121, 131
   rimu 33
   see also grapeweed, bloodweed
weka 152
welcome swallow 68
whale
   blue 79
   Bryde's 28, 36, 44, 52, 79
   false killer 23
   fin 79
   humpback 23, 36, 79, 84, 89, 100
   minke 23, 36, 79
   orca (killer whale) 23, 28, 36, 44, 52, 73, 79, 84, 89, 94, 100, 116, 122, 157
   pilot 23, 84
   sei 23, 36, 79
   southern right 84, 89, 100, 129, 134
   sperm 23, 89, 134, 157
Whanganui Inlet 102–4
   whelk
   lined 56–7
   mud 62, 67, 104, 111
   northern siphon 67
   spotted 56–7, 67, 105
whitebait 104–5, 111
white-capped noddy 23
wrasse
   banded 27, 42, 50, 73, 84, 89, 93, 99, 110, 116, 121, 126, 131, 133, 150
   elegant 22, 34
   girdled 126, 132–3, 150, 157
   green 22, 27, 44
   orange 22, 33
   Sandager's 22, 27, 34, 44, 73, 79
   scarlet 27, 44, 51, 73, 79, 84, 89, 94, 126, 133, 150
wrybill 62, 68

yellowhammer 45

zoanthid 35, 84, 133